Ocean Governance (Beyond) Borders

Kimberley Peters · Jennifer Turner
Editors

Ocean Governance (Beyond) Borders

Editors
Kimberley Peters
Marine Governance, Helmholtz Institute for Functional Marine Biodiversity
University of Oldenburg
Oldenburg, Germany

Jennifer Turner
Cultural and Political Geography
Universität Trier
Trier, Germany

ISBN 978-3-031-71321-7 ISBN 978-3-031-71322-4 (eBook)
https://doi.org/10.1007/978-3-031-71322-4

Cover illustration: AerialPerspective Images

This Palgrave Macmillan imprint is published by the registered company Springer Nature Switzerland AG
The registered company address is: Gewerbestrasse 11, 6330 Cham, Switzerland

If disposing of this product, please recycle the paper.

Acknowledgements

This book started as a set of conference sessions entitled *Ocean Governance Beyond (and Against) Borders* at the 2021 Royal Geographical Society (with Institute of British Geographers) Annual International Conference in London. We thank the contributors of those sessions, some of whose chapters appear here, but all of whom we are grateful to have worked with and thought alongside. Some chapters found their way to the project later, and we thank the authors who agreed to join this discussion in the pages to follow. It was after those initial conference sessions that the idea of this book was first floated with Rachael Ballard. We owe her and the team at Palgrave (including Hemapriya Eswanth, Zeenathul Raeesa Ismail and Connie Li) much thanks in their support as the project unfolded. We are also grateful to those with whom we discussed the concepts and ideas of this book, including notably the Marine Governance Group at HIFMB. This book is Open Access with the support of the Alfred Wegener Institute—Helmholtz Centre for Polar and Marine Research, with thanks to Andreas Walker, Head Librarian and Tore Hoffmann of Library Services especially.

This was a book that developed over the past three years. In that time there have been significant changes to ocean borders: social, cultural, political and environmental (with overlaps of these too). Such borders—related to people and planet—will continue to change into the future (not least with a new, major Agreement under the United Nations Convention on the Law of the Sea on the Sustainable Use of Marine Biological Diversity of Areas beyond National Jurisdiction (the so-called BBNJ Agreement) being adopted in 2023, which includes statements on Area Based Management Tools (ABMTs)—the creation of bounded ocean zones). As such, this book is both complete, and incomplete. It should be read as a provocation to think more on ocean borders, and to encourage further conversation as more borders materialise in the oceans, as they undoubtedly will.

In addition to ocean borders, the boundaries marking our timeline for this book stretched several times. Yet this collection was a project that needed time and care. We are thankful for the kindness that has been part of this process, and are grateful to have once again worked with one another—and to have made it over the line.

Contents

Notes on Contributors

Joseph Onwona (Kofi) Ansong is an expert coastal and marine planner. His career has focused on critically examining the challenges and opportunities for marine spatial planning as well as its technical development. His work has brought together scientists, policymakers, regulators and industry to find solutions for transboundary marine governance and delivering a sustainable blue economy. Kofi is a member of the UNESCO-IOC MSPglobal Expert Group, the MSP Research Network Steering Group and an Honorary Visiting Scholar at the University of Liverpool, UK.

Maria Borovnik is a geographer and Senior Lecturer in Development Studies at Massey University, New Zealand, working in the areas of mobilities, migration and development studies with a special interest in mobile occupations. She co-coordinates the Mobilities Network for Aotearoa and the New Zealand Geographical Society's Mobilities in Geography Research and Study Group and is on the Editorial Board of *Transfers: Interdisciplinary Journal of Mobility Studies*. Maria co-edited *Weather: Spaces, Mobilities and Affects* (with Barry and Edensor, 2021) and is currently working on the Royal Society of New Zealand Te Apārangi funded research project 'Navigating labour mobilities: Seafarers after COVID-19'.

Nancy Couling is a researching architect at the ETH Zurich and Associate Professor at the Bergen School of Architecture, inspired by intangible, invisible yet critical large-scale urban phenomena, in particular the urbanisation of the sea. Following architectural studies in Aotearoa/New Zealand, she was awarded a postgraduate fellowship at IUAV, and worked in international practices before co-founding her own interdisciplinary agency cet-0/01 in Berlin. Her Ph.D. was awarded at the EPFL (CH) in 2015, and her postdoctoral Marie Curie Fellowship was carried out at the TU Delft 2017-19 with Carola Hein, published in 'The Urbanisation of the Sea: from Concepts and Analysis to Design' (Couling & Hein, 2020, Rotterdam:nai010).

Oscar Hartman Davies is a Social Sciences Engagement fellow at the School of Geography and the Environment at the University of Oxford and a visiting researcher at the KTH Royal Institute of Technology in Stockholm. His work centres around the interdisciplinary field of digital ecologies, and his Ph.D. explored the mobilisation of seabirds as sentinels of environmental change and their entanglement with digital transformations in fisheries governance. Oscar is also an advisor to several youth-focused nature recovery initiatives in the UK and Europe.

Po-Yi Hung is a professor of geography at National Taiwan University. He uses food, agriculture, and fisheries as lens to look into border and territoriality, mobility and infrastructure, nature and society. He has conducted research in Taiwan, China, the Highlands of Southeast Asia, and Mauritius in the Indian Ocean. Po-Yi is the author of *Tea Production, Land Use Politics, and Ethnic Minorities: Struggling over Dilemmas in China's Southwest Frontier* (2015, Palgrave Macmillan). His latest studies focus on two themes: one is about water and tea production under climate change, and the other is the geopolitical ecology of fisheries.

Yvonne Kunz trained as a human geographer, is currently Acting Professor for Sustainable Spatial Development and Governance at the University of Trier, researching the acceptance and regulatory frameworks of (potential) vegetation-based Blue Carbon initiatives. Prior to this engagement, she worked at the Royal Netherlands Institute of Southeast Asian and Caribbean Studies (KITLV) where she coordinated a trans-disciplinary project on Marine Protected Area governance in a changing climate. Her work focuses on natural resource regulations for land, air and water, mainly in Indonesia and the Philippines, recently also in the Dutch Caribbean.

Christopher McAteer is currently a Ph.D. candidate on the Social and Political Thought programme at York University, Canada. His dissertation examines the challenges for social and political theory in comprehending the temporal complexity of climate change. The project engages with international relations theory, political theory, decolonial theory, aesthetics, and music to formulate a heterotemporal theory of the world-political present of climate change. He is also a composer represented by the Contemporary Music Centre, Dublin.

Jasper Montana is a senior lecturer in Science, Society and Environment at the Centre for the Public Awareness of Science at the Australian National University and an honorary research associate at the School of Geography and the Environment at the University of Oxford. His work examines the politics of nature conservation and the geographies of knowledge in environmental governance. Jasper previously completed a Leverhulme Early Career Fellowship and Ph.D. exploring the production and use of knowledge in biodiversity conservation.

Kimberley Peters leads the Marine Governance Group at the Helmholtz Institute for Functional Marine Biodiversity (HIFMB). Kim researches how watery spaces are managed for environmental protection. She is a socio-cultural and political geographer training and has completed projects focused on marine protected areas, deep-sea mining and ship routeing measures. Kim is the author/editor of 8 books including: *Water Worlds: Human Geographies of the Ocean* (Ashgate, 2014), *Territory Beyond Terra* (Rowman and Littlefield, 2018), the textbook *Your*

Human Geography Dissertation (Sage, 2017) and the essential social science collection on maritime worlds: *The Routledge Handbook of Ocean Space* (Routledge, 2022).

Katherine G. Sammler is an assistant professor of Environmental Knowledge, Technology, & Sustainability Studies at the University of Twente, Netherlands. Katherine is a geographer using feminist science and technology studies, more-than-human political ecology, and critical geopolitics to analyse entangled nature-culture relations. Their research focuses on the convergence of science and politics through knowledge production and material practices in ocean realms and outer space. Recent publications include theorisations of digital automation in seabed mining, the politics of gravity in managing bodies and waste in orbit and the role of sea level metrics in defining global volumes.

Lara Şarlak is a Ph.D. candidate at the University of British Columbia's Department of Anthropology, Canada, exploring the essential role of migrant recycling workers in supporting the urban waste management infrastructures of their hometown Istanbul, Turkey. Lara received their M.A. in the Anthropology of Media at SOAS University of London and a B.A. in Sociology and Media and Visual Arts at Koç University in Istanbul. As a visual anthropologist, Lara strives to incorporate their audio-visual skills into ethnography in a publicly accessible way. Lara's active involvement in community-driven initiatives providing support to migrants, women and LGBTQI+ people in Turkey also profoundly informs their research.

Satya Savitzky worked as a research scientist at the Helmholtz Institute for Functional Marine Biodiversity, Germany. Prior to this he was a post-doctoral fellow at St Andrews University, UK. He completed his thesis, 'Icy Futures: Carving the Northern Sea Route', at Lancaster University in 2016. He has published several peer-reviewed articles and book chapters examining the creation, maintenance and contestation of routes, and the politics of governing mobilities, on land and at sea. He was the co-editor of the book *Cargo Mobilities* (Routledge, 2016). His novel research was most recently examining ships (and ballast tanks) as spaces of turbulent ecologic circuits.

Jennifer Turner is a human geographer with a strong interest in societal borders and boundaries. This interest first developed in the context of the boundary between prison and society, attending to the blurred relationships between the 'inside' and 'outside' of carceral spaces through examples including prison architecture, staff employment, prisoner work programmes and the prison-military complex. In developing conceptualisations of the 'carceral', Jennifer co-edited *Carceral Worlds: Legacies, Textures, Futures* (with Stuit and Weegels, Bloomsbury, 2024) and focuses on other non-prison spaces that may operate under similar conditions of spatial control and detriment, such as carceral seas.

Julia Verne is a chair of cultural geography at the Johannes Gutenberg University Mainz. Her long-term ethnographic fieldwork among Zanzibari-Omani traders has turned her attention to the relevance of the Indian Ocean as a specific material and imaginative entity and led her towards maritime geographies. Inspired by STS, posthumanist approaches and an interest in the ways in which technologies mediate human-world relations, she has recently published on the role of digital technologies in East African agriculture, wildlife management and disaster communication, as well as on more general questions regarding geographic engagements with the Anthropocene.

List of Figures

Counter-Mapping: A Morphology of Oscillating Margins in the Norwegian Sea

Contested Borders and Resolution in Planning Shared Marine Waters

Can Borders in the Ocean Respond to Climate Change?

Introduction: Closures—Ocean Governance Borders

Kimberley Peters and Jennifer Turner

Closures: An Opening?

This book opens with closures: border closures, and those occurring in the planetary spaces often understood to be the most open—the seas and oceans. The Weddell Sea, situated in the Southern Ocean surrounding Antarctica, for example, appears to be the last 'pristine ecosystem'—an open, icy, expanse of space (Teschke et al., 2021: 3). However, it is simultaneously one that has been subject to significant scientific monitoring and subsequent line-drawing exercises in efforts to enclose 'domains' for marine protection (CCAMLR, 2013; see also Teschke et al., 2021). Whilst the establishment of a network of marine protected areas (MPAs)

K. Peters (✉)
Marine Governance, Helmholtz Institute for Functional Marine Biodiversity at the University of Oldenburg, Oldenburg, Germany
e-mail: kimberley.peters@hifmb.de

J. Turner
Cultural and Political Geography, Universität Trier, Trier, Germany
e-mail: turner@uni-trier.de

K. Peters and J. Turner (eds.), *Ocean Governance (Beyond) Borders*,
https://doi.org/10.1007/978-3-031-71322-4_1

is desired by those seeking to preserve the area against a range of anthropocenic pressures, the borders of a Weddell Sea MPA specifically have yet to be agreed and enacted by nations party to CCAMLR (the Commission for the Convention on Antarctic Marine Living Resources), the body which has authority in allocating such areas of protection. What such attempts demonstrate, however, is the presence of ocean borders for biodiversity protection—the use of lines to potentially *govern* and control space and enclose it into what Noella Gray calls 'conservation territories' (2018). Indeed, governance refers to '"the art or practice of governing" but is now a more complex term' (Johnston et al., 2002: 317), whereby it relates to 'any … mode of co-ordination in which the aim is to control, guide, facilitate economic and social activities … including activities involved in transforming nature' (Jonas & While, 2005: 72).

Indeed, far from being an innocent exercise of spatial demarcation, MPAs are a practice of governing that create bordered units of space, 'functional enclosures' that are underscored with particular values on how space should be used (or not) (Lambach, 2021). They may work in the interests of some, and not others (conservationists compared to fishers, for example). The politics of MPA designation, and more so the complexities in values underlying them, are well examined (including in this book). Nonetheless, MPAs are typically designed to replenish and 'restore'[1] the oceans, their more-than-human abundance, species richness and other biodiversity traits. As a report from the International Union for the Conservation of Nature (IUCN) writes, 'MPAs aim to protect all the features of importance *within their boundaries*' (Day et al., 2019: 16, emphasis added). The borders they rely on create a governance regime that operates *against* particular uses, or by setting limitations in use (for example restricting ocean access to bounded zones for license holders only, setting fishing quotas, or allowing use only at particular times of year). In the case of MPAs, then, borders work in multiple ways,

[1] Critical scholars, including political ecologists and human geographers have noted the politics of restoration, where such acts of conserving rest of decisions as to what to conserve (and what to forgo) in terms of environments, species etc. (see Adams, 2013, 2020), as well as noting the temporal dimensions where habitats and ecosystems can never truly be restored to a former state, but where, through the passage of time they emerge in a new state altogether, through their conservation, whilst being haunted by pasts (McCorristine & Adams, 2020).

creating an 'inside'—a bordered space of rules and order to protect the natural world (in some visions and not others, see Peters, 2020). This functions by locking *out* other activities such as trawl fishing, leisure and tourism, or highly regulating them. Borders then, are a result of drives for ocean governance—their enactment is a consequence of approaches to governing. Yet, once in place, MPAs then act as ocean-governing instruments themselves, b/ordering the oceans in specific ways.

In a rather different but no less pertinent example, the UK's Conservative government's 'Stop the Boats' campaign (notably active throughout 2023–4 in an effort to curb unauthorised travel to the UK via the English Channel) represented a hard*line* strategy in solidifying the oceanic border of the nation. The slogan of the campaign rested on an oft-tried strategy of stretching border defences out to sea to stop the arrival of migrants and asylum seekers before they ever reach the shore (see Mountz, 2020; Mountz & Hiemstra, 2012). Indeed, the territorial border was to be governed—though a variety of practices—including surveillance and then the interception of small boats carrying people seeking asylum, travelling through 'unofficial' routes. Whilst attempts to limit and halt the arrival of asylum seekers sees strategies also centred on other spaces including the land and also air (including, in the UK context, the threat of further removal of arrivals to destinations such as Rwanda), the 'Stop the Boats' campaign was a directly *oceanic* intervention of increased security via the patrol of the Channel separating (and connecting) Britain and/from France. In this relatively small section of water, only 23 nautical miles across, border regimes played (and continue to play) out in the sea, 'defending' the invisible, watery, territorial line of the nation state—'protecting' the 'inside' of the nation state from the 'outside' and 'outsiders'. As Henry Jones remarks, '[l]ines are never just on maps. It is when people encounter the force of a border that they experience the violence that is needed to make those lines' (2016: 343). Similar border regimes have been seen to play out at sea globally, as migration policies shift offshore in seas and oceans as diverse as the Caribbean Sea, Mediterranean and offshore Australia (see Loyd et al., 2016; Stierl, 2021; Dickson, 2021 respectively).

Across the oceans, then—even if they may be imperceptible to the eyes—lines are etched that draw borders for a variety of diverse reasons but, notably, in acts of control. As Philip Steinberg (2001) has noted, the oceans have been *constructed* through human acts of demarcating, allocating, claiming and regulating space, for differing reasons, at different times—often through definitional line -drawing exercises (see also Steinberg, 1999). The diversity of placing lines—or borders—in the ocean, includes lines that delineate ocean zones (such as coastal nations' territorial seas from their Exclusive Economic Zone (EEZ) to the High Seas [Vadrot, 2020]); lines that bound zones of marine conservation (Gray, 2018; Lambach, 2021); or lines that allocate ocean space for specific uses (such as shipping routes, energy development, or extractive industries) under regimes of marine spatial planning (Jay, 2010, 2020). As Steinberg (1999: 261–2) notes, across the seemingly monotonous plain of world oceans, there exists a manifold of lines not forgetting of course 'equatorial lines, latitude and longitude lines, rhumb lines, and bathymetric lines'.

In writing of maritime borders, Po-Yi Hung and Yu-Hsiu Lien (2022: 870) explain that 'it is not an exaggeration to say that, in reality, borders are everywhere'. Indeed, far from being empty, free of governance, or a watery 'wild west' (Langewiesche, 2010; Urbina, 2019) oceans are full—spilling over even—with border *practices* which govern their use, with those borders in turn then becoming tools of governing movements, resources, people, and so on. Yet it is important to note that borders are not only lines that exist (on maps, if not on the ocean surface itself, a space that defies inscription [see Steinberg & Peters, 2015]). Those border lines are maintained, secured, transgressed and challenged through a variety of *performances* of/at the border—at its edge and far beyond it (see Paasi, 2013, for a review). If we follow Hung and Lien's assertion, aside from the two examples we gave at the start of this chapter, there are many other ways in which borders are present in, and placed within, the oceans. And, subsequently, there are many more humans (in addition to migrants, enforcement officers, fishers) and more-than-humans (in excess of a diversity of fish and mammals) enacting and experiencing how oceanic borders (are put to) work. Such recognition of the extent of bordering, and the range of oceanic actors that form part of these processes, is reflected in further examples that are featured in the following chapters of this book. These include, for example, reflections on the dynamics of MPAs and the policing of asylum movements across the ocean (as two archetypal examples), but chapters also cover shark management techniques for leisure and tourism purposes, the border regimes impacting mariners who are part of global shipping fleets, the microorganisms travelling across political and ecological borders in ship ballast water, and more.

The book began as a set of conference sessions at the 2021 Royal Geographical Society (with Institute of British Geographers) Annual International Conference during the Covid-19 pandemic, where participants were physically separated by the newly enacted border controls on travel, designed to prevent contagion. Here, the editors convened two online sessions mirroring the title of this book. The aim of the sessions was to reflect on one of the planet's most boundless spaces, at a moment at which the sea was appearing to become more 'carceral' (Moran et al., 2018)—increasingly contained and confined through bordering processes. This was most evident in that moment through border regimes which held seafarers indefinitely offshore due to the industry vector risks of coronavirus spreading across borders. As "Borders and Confinement in Seafarers' Realities" (Maria Borovnik) reveals, this bordering would cause many harms to those living and working at sea who continued to service the global economy but were often not protected as 'key workers'. Some chapters of the book have been developed from the original set of sessions. Others came to be included through disciplinary dialogue on ocean borders as the book developed, and as discussion on borders and the oceans continued to grow (see Hung & Lien, 2022). What results here is a first book that shifts across varying oceanic terrains to offer, in one collection, a sustained reflection on bordering and oceans, written by those researching and reflecting on such debates.

We started this chapter with the opening two examples (conservation bordering and asylum control) because they share conceptual similarities—and ones that run throughout the book. They demonstrate how borders, emplaced in the ocean, work to locate and allocate spaces for specific uses, for specific people (and not others), through a spatial politics of containment or enclosure. That is not to say crossings do not occur, but ocean borders are enacted to try and control a medium whose mobile churn seems to deny/defy modes of b/ordering (Steinberg &

Peters, 2015). In spite of the sea's resistance, humans are resilient in their efforts to draw lines. Accordingly, in this introduction we consider *closures*: providing a conceptual underpinning for a book that presents a variety of windows into ways in which the ocean, and the various human and more-than-human lives entangled with it, exceed (Peters & Steinberg, 2019) attempts at bordering. The book to follow, then, considers the ways borders are planned, enacted, monitored and enforced but also the ways in which they are porous, leaky, subject to change, transgressed and contested, all in a largely mobile medium that is itself resistant to bordering (Peters, 2020). We begin in the next section by zooming out to review some of the pertinent longstanding literature on borders, boundaries and enclosures, before, in the section thereafter, examining such ideas in the particular context of the seas and oceans. We use our discussion of (en)closures as an *opening*, to introduce the key themes that weave through the chapters, as well as summarising those the chapters briefly. We close out this chapter with ideas to frame readers' engagement with the book, and some thoughts on how the contributions herein might be navigated.

Understanding Borders

As highlighted already in this introduction, borders rest on a particular dynamic: the creation of insides and outsides (Turner, 2016). As is evident in either of the two previous examples of conservation approaches or migration strategies, this applies to at sea as much as it might on land, where an array of materialities (walls, fences, etc.) technologies (cameras, (remote) sensing devices, and so), and processes (the administrative functions of bordering) govern areas or units of *grounded* space, creating zones of regulation within, and different rules and regimes, *beyond*. Borders and processes of spatial control are also enacted in other planetary terrains beyond the land, including the sea (Peters et al., 2018) and even outer space (Beery, 2016). Yet this central idea of borders as spatial dividers—making insides and outsides—fits with the conventional definition of borders as "strategic lines" (often in relation to military endeavour and the defence of lines of combat

[Turner, 2016: 30]). As Reece Jones (2009: 180) notes, borders tend to refer to *political* units (as opposed to the term 'boundary' which might have a wider application). Traditionally borders and acts of bordering are about defining political spaces for a given purpose (often control of geographical areas, peoples, resources etc.) creating what Stuart Elden calls 'bordered power container(s)' (2010: 799) or *territories.* These containers or territories function through regimes of controlling what is within and, in doing so, also establishing what is beyond as *outside of*, *or other to*, what is inside. However, whilst borders have developed a particular association with national territories and other political zones within them, the term border also has a wider resonance. Political borders are not just national ones and they may be enacted in any number of spaces, including the ocean (Hung & Lien, 2022).

Moreover, borders are more than lines, but points of crossing and exchange. As Anssi Paasi notes, borders are not 'separate sociospatial entities' (2012: 2304)—they do not operate in isolation: simple, singular lines in the land, or ocean, or elsewhere. They are relational in that they separate and *connect*. Indeed, Turner (2016) stresses that the border is not a hard and fast line of socio-cultural and political separation. Think, for example of the border between land and sea. There is no neat division between these spaces (Ryan, 2012). As Hung & Lien note (2022: 882–3) '[t]he "in-betweenness" of tidal flats makes them a unique space that is neither terrestrial nor oceanic, and which both separates and connects land and ocean … Accordingly, it is important to ask how thinking about maritime borders will go about connecting land and ocean'. The intertidal zone shows that there is no clear line delineating water from 'earth'. Rather, land-sea divisions (like so many) are slippery (Choi, 2022) and shifting (Jones, 2011). As Uma Kothari (2020: 64) writes, '[t]he land-sea border is constantly ebbing and flowing, merging the geophysical spaces of sea and shore … land and sea, so often envisaged as definitively delineated yet increasingly acknowledged as fluid, highlights the extent to which these boundaries are difficult to identify, often blurred and incessantly changing'. Indeed, borders can be thought of, in more recent literature, not as hard and fast lines but as points of crossing (Paasi, 2013; Turner, 2016), exchange (Newman, 2006) or mixing (Paasi,

2016). In this sense, the land-sea border exemplifies ideas about border dynamics per se.

This is not to say that the hardening of borders is not occurring, with tools and approaches aimed at preventing infiltrations, mergings and other cross- and through- border relations. Border 'strengthening' and 'defence' are frequent political strategies, as well as, correspondingly, core academic concerns (R. Jones, 2016). Yet borders are made—but also function—through multiple relations and power dynamics that span across borders (what Turner [2016: 230] also calls a 'patchwork' of processes that make borders function as they do, as sites of blurring, fuzziness, exchange and crossing). As Johnson et al. additionally note, '[m]uch... bordering work... happens far from the political border' (2011: 61) where borders are made, re-made and contested *at distance*. For example, regimes of border management at sea may be made on land (such as the provisions of the United Nations Convention on the Law of the Sea (UNCLOS) which sets out the territorial sea limit—a limit made in the headquarters of the UN, but then practiced in the seas and oceans themselves).

And, moreover, borders are not just spatially diffuse (i.e. not static in space), they are also not static in time. Borders are temporal in that they change, adapt, and are reworked: through official and unofficial means, via political gerrymandering and acts of resistance. As such, 'borders are ... historically contingent ... and constitutive of the perpetual production and reproduction of territories': they are contextual, and constantly made and re-made (Paasi, 2012, 2304). They are also, then, not ahistorical: borders tell stories about the governance of space, resources, life, over time. To this end, borders are also both 'material and immaterial' (Turner, 2016: 30). They are physical but also have other qualities, with the temporal element being just one. Borders are also social. Indeed, borders do not just establish lines in space, they establish regimes, norms, and senses of belonging—for human and more-than-human life.

Borders, as we saw at the start, define the rules within specific spatialities—whether this is via legal regulations and even unwritten social rules—and hence outline which people, and what activities and behaviours are acceptable, or not. For example, the MPA border is, if not a physical line in the ocean, a definitive line on a map that draws a material area of protection. But the border also defines expectations, and requirements: what is appropriate (and not), and to what degree (in terms of fishing, and other resource extraction activities, tourism and scientific endeavour). The border is therefore also a social valuing device—stressing what matters and what should be 'saved'. The borders of MPAs manifest materially (in paper regulations and on electronic maps), but they are also immaterial in that they refer to the very ideological basis that underscores spatial control. Likewise, surveillance and policing of borders in view of those seeking asylum through watery crossings also represents both the defence of the material border of the nation state—the line marking British territory—and notions and ideas of who belongs where, who should be allowed within a territory (for what reasons) or not. As Newman neatly summarises (2006: 143), '[w]e live in a world of lines and compartments. We may not necessarily see the lines, but they order our daily life practices, strengthening our belonging to, and identity with, places and groups, while—at one and the same time—perpetuating and reperpetuating notions of difference and othering'. Borders then, are social *and* spatial (see also Cresswell, 1992, 2014).

Generally speaking, then, border studies is a complex and multifaceted field that has considered the material qualities and performative practices of borders across a variety of examples as well as *conceptualising* the border and its workings. As Hung and Lien (2022: 881) deftly outline, border studies have 'progressed via a series of different "turns," including the "processual turn," (Brambilla, 2015; Yuval-Davis et al., 2019), the "mobility turn" (Salter, 2013; Steinberg, 2009), and the "material turn" (Sur, 2013)'. Moreover, border studies enrols different disciplines (from geography, to political science and International Relations and beyond) in its attempts to make sense of articulations and operations of borders. As Reece Jones notes (2009: 184), '[d]espite the fundamental role which boundaries and categories play in ordering, organizing and limiting everyday life, there is not a singular field of study that investigates them'. This book brings together a cross-border collection of authors who use some of the ideas about borders noted here (borders as spatial, temporal, fixed, porous, material, social, conflictual etc.) to examples at sea. The book traverses disciplines as wide as anthropology

(Lara Şarlak) and architecture, (Nancy Couling), to Marine Spatial Planning (Joseph Onwona *(Kofi)* Ansong), political ecology (Katherine Sammler, writing with Satya Savitzky and Kimberley Peters), as well as human geography (Maria Borovnik, Po-Yi Hung, Yvonne Kunz, Jasper Montana and Oscar Davies, Julia Verne) and social and political thought (Chris McAteer). As Paasi writes (2012: 2304), '[c]urrent interdisciplinary research on borders varies both thematically and in its theoretical, methodological and empirical orientation'. It is the 'empirical orientation' towards the seas and oceans, which we consider next.

Borders and the Ocean

The oceans are facing unprecedented rates of *enclosure* through various regimes of bordering, with Luke Fairbanks and colleagues (2018: 148) noting that the oceans are currently entering a 'third phase' of enclosure. Here the authors refer to three major, global acts of oceanic bordering: first the creation of a territorial limit for national waters (set to 12 nautical miles); second, the making of a defined Exclusive Economic Zone for coastal nations (up to 200 nautical miles); and thirdly, an intensification of bordering of oceans through conservation practice that increasingly happens in the High Seas (Gray, 2018). Indeed, Fairbanks et al. note that '[s]tates began to limit … freedom in the late nineteenth century with modest expansions of sovereignty over near coastal seas' (resulting in 12 nautical mile delimitation of the third version of the United Nations Convention on the Law of the Sea) (2018: 148). The authors note that the seas have been further 'territorialized through EEZs, fixed to appropriate resources' (ibid: 148), creating what Campling and Havice have described as 'the largest single enclosure (of space) in history' (2014: 714). Now '[p]roperty rights … and conservation efforts that value, commodify or protect marine ecosystems' represent the 'third phase' of major, global bordering efforts (Fairbanks et al., 2018: 148). As Gray notes 'conservation territories are spatial interventions premised on legal and/or other institutional systems that rework human–environment relations and resource access and control in particular ways' (Gray, 2018: 257).

Yet this current legal regime in which major ocean lines are drawn masks a much longer history of political bordering in the seas and oceans. Indeed, Philip Steinberg has noted that the grounding sentiments within the United Nations Convention on the Law of the Sea (UNCLOS), which sets out these divisions of ocean space, are far from recent, but rather the consequence of the 'institutionalisation of an ocean-space construction developed over the previous 200 years' (Steinberg, 2001: 149).[2] Steinberg has demonstrated that ocean lines and political bordering have been constants in the ocean, not least through various frames of capitalism as the ocean has been variously *constructed* as a resource for humans (see Steinberg, 2001). That political borders at sea are also linked to the nation state and its operation(s) also reveals the foundation on which current 'phases' of ocean enclosure are based: one that is European and Western focused, where the 'settler-colonial governmentality of natural resources' dominates modes of 'management' and where the planet is reduced to 'resource', rather than 'force' (George & Wiebe, 2020: 500). Indeed, Lambach (2021: n.p.) notes how ocean bordering is framed within 'prevailing late modern notions of legibility, control' rather than 'some greater effectiveness of spatial instruments' such as lines and demarcations. Peters has also shown how bordering has a territorial and landed logic, and has been 'transported' to the sea, making borders a 'go to' strategy in control and governance but one that is 'so powerful it is rarely questioned' (Peters, 2020: 4).

Indeed, that the seas and oceans are subject to bordering, and how they are subject to bordering, is a key topic of this book. As Po-Yi Hung and Yu-Hsiu Lien note (2022: 870), 'current research has mostly concentrated on issues concerning terrestrial borders'. There is a need to explore in greater depth, 'how the seemingly boundless oceans are in actuality subject to a variety of bordering forces' (Hung & Lien, 2022: 870). In Hung and Lien's agenda-setting paper, they set out a variety of ways to examine ocean borders, and this book follows the lead in offering a comprehensive consideration of modes of bordering from the control of resources, of people, to control for conservation

[2] With 'conservation territories' in the High Seas (Gray 2018) now part of the amendment adopted in 2023 – the BBNJ agreement – the management of biodiversity in areas beyond national jurisdiction.

management and all against the grain of the ocean's geophysicality (see also Peters, 2014; Peters et al., 2018; Peters & Steinberg, 2019; Steinberg & Peters, 2015). As Kothari writes, the ocean presents a paradox when it comes to bordering 'as a realm that is incessantly shaped through relational processes that transcend finiteness and boundedness' (Kothari, 2020: 66). As Braverman also puts it, borders 'run against, or intersect, traverse and conflict with the chaotic, fluid, deep spatiality of the oceans' (Braverman, 2022: 5).

The seas and oceans, are, for the most part, fluid, mobile, liquid and churning (Steinberg & Peters, 2015). They are also much more than this. They can be solid, as ice, gaseous as water evaporates, as well as imaginary existing in creative constructions in film, books, poetry and so on (see Peters & Steinberg, 2019). Yet, as a predominantly watery, wet, mass, they can be understood to defy inscription (Schmitt in Steinberg & Peters, 2015: 249). It presents a challenge to make physical markings in the shape of ocean lines, to build fences or erect other forms of tangible borders. Even immaterial ones may be challenging—such as surveillance practices or Global Positioning System (GPS) tracking, over an expanse as vast as the many seas and oceans that constitute our world (see Peters, 2014 for the challenges of oceanic surveillance, also Nyman, 2019). Although various actors, from state agencies to business may wish to construct borders, the ocean's materiality frequently and stubbornly resists such efforts. As Peters (2014) demonstrated in a study of enforcement measures waged against offshore pirate radio stations in the 1970s, it was difficult for British authorities to prosecute pirates. They could only do so when ships were *inside* their territory, in territorial waters where their law applied. Beyond this line, they had no jurisdiction. Yet, accurately defining the line dividing British territory from the High Seas beyond was problematic (not least at a time without enhanced technology to log and track vessels). The line—the border—was not physically marked. It was difficult to obtain proof. Was the vessel inside or outside?

Scholars have noted that whilst '[z]ones, areas, sectors. Borders, boundaries, limits… (are) quite often, the language deployed in relation to ocean governance' (Peters, 2020: 1) there are special challenges of how these function in wet space. As Sara Maxwell and colleagues have noted of marine protection borders: '[m]ost spatial marine management techniques (e.g., marine protected areas) draw stationary boundaries around often mobile marine features, animals, or resource users. While these approaches can work for relatively stationary marine resources, to be most effective marine management must be as fluid in space and time as the resources and users we aim to manage' (Maxwell et al., 2015: 42). Here Maxwell et al. acknowledge that borders run counter to the geophysical logics of the water and species within—which move across, through and beyond the borders of ocean management. Put another way,

> The ocean… through its material reformation, mobile churning, and nonlinear temporality—creates the need for new understandings of mapping and representing; living and knowing; governing and resisting. Like the ocean itself, maritime [and marine] subjects and objects can move across, fold into, and emerge out of water in unrecognised and unanticipated ways. It is in this context that we advocate thinking from the ocean as a means toward unearthing [new modes of governance tools] … (Steinberg & Peters, 2015: 261–262)

This book explores this tension: of governing a mobile, wet, space. But it also acknowledges that these tensions run deeper. The sea is also a space where bordering tends to happen surficially. However, the sea is voluminous and deep, and bordering relates often to what is *below* even if governance is most often articulated through horizontal demarcation. As Hung and Lien write (2022: 875) 'while the marine surface is of definite importance, the ocean is much more than just a surface'. The chapters to follow attend to the ocean in its various forms: as theory machine (McAteer), as imagined (Montana and Davies), as materially violent (Sarlak), as subject to wider forces, such as climate (Kunz). It is to these chapters we turn next, outlining how they take on the topic of borders, before setting out how readers may wish to engage with the contents of this book.

Navigating the Way Ahead

The Contributions

This book offers multiple, diverse perspectives on ocean borders, with each contribution seeking to both describe but also examine how and why borders are put to work in the oceans and to what end. At a moment where ocean governance is a 'hot' topic among academics, policymakers, governments and non-governmental agencies alike, what follows takes on one of the most overlooked but central devices underscoring many modes of oceanic management: the border. Combining contemporary border scholarship with cutting-edge ocean governance research, this edited volume explores varied bordering practices but also aims to challenge the ontological stability or 'common-senseness' with which bordering is deployed in seas and oceans. Throughout the ten carefully authored chapters that follow, together with this introduction and conclusion, the book asks: *What borders are present in the seas and oceans, where and why?* In doing this, it further considers: *Do we need borders? Can we govern differently? What would ocean governance that goes beyond bordering practices look like, entail and enable?*

Throughout the book, from varying disciplinary perspectives, the logics of bordering are tested, before investigating how such logics relate to people and impact the planet. For people, maritime borders have historically, and in the present, resulted in devastating violences (Khalili, 2021). One such example is via dimensions of oceanic migration and offshore detention where borders and buffer zones are enacted (Kothari, 2020; Stierl, 2021). For the planet, marine borders have been uncovered as insufficient for governing mobile worlds where marine life, and interactions with it, move (Maxwell et al., 2015; Peters, 2020). Borders have also been associated with neocolonial ocean 'grabs' (Bennett et al., 2015; Foley & Mather, 2019). As such, bordering can be deeply problematic. Yet, it persists. This book shines light on that persistence and its a/effects.

This volume presents a wide-ranging analysis of bordering practices, bringing together a unique collection of chapters—crossing the boundaries of social science and the arts and humanities, as well as policy and practice—to take seriously the work of bordering at sea, whilst also, fundamentally, interrogating its very basis. The chapters reflect on the bordering processes in ways that combine theoretical engagement with empirical rigour, each offering a different angle into the topic of ocean governance (beyond) borders. In "Overdetermined by Territory? Governing the Ocean in Time, Matter, and Rhythm", Christopher McAteer takes the angle of **time and materiality** in relation to borders (where space has been traditionally dominant). In "Counter-mapping: A Morphology of Oscillating Margins in the Norwegian Sea" Nancy Couling approaches borders by engaging ideas of **vertically and volume**, exploring counter-maps of the ocean that engage its spatiality and temporality more fully. Po-Yi Hung deploys the lens of **territory and geopolitics** in "Bordered-In, Bordered-Out, and Overlapping Territorialities in Ocean Space: The Case of Fisheries" to think about borders and power, using the example of fisheries. In "Contested Borders and Resolution in Planning Shared Marine Waters" Joseph Onwona *(Kofi)* Ansong continues to think with power to survey the role of **nations and planning** efforts in delineating contested borders in organizing use of shared marine waters. "Imaginaries: Oceanic Bordering with Large-Scale Marine Protected Areas" turns from a material, deep, mobile ocean territory, to one where Jasper Montana and Oscar Davies explore how **imaginations and governance** intersect in the establishment of large-scale marine protected areas. Continuing this theme, Yvonne Kunz asks, "Can Borders in the Ocean Respond to Climate Change?" Here, Kunz considers if bordering ocean zones can respond to the more elusive yet expansive impact of climate change by considering the **discourses and practices** driving conservation.

From the ocean to the life within it, Satya Savitzky, Katherine Sammler and Kimberley Peters offer "Bordering Marine Belonging: The Meanings, Mobilities and Materialities of Bioinvasion" as a consideration of questions of **space and security**—how certain species are deemed to belong somewhere and not elsewhere, and are hence framed as biosecurity and invasion 'risks' and subject to bordering attempts of management. Continuing to think about the more-than-human world, Julia Verne, in "Human-Shark Encounters Beyond Borders: (Post-humanist) Attempts to Navigate a Maritime Contact Zone", weaves a discussion of the relations between **life and technologies** in touristic management of the threat of shark encounters. Maria Borovnik ("Borders and Confinement in Seafarers' Realities") shifts attention back firmly to people and focuses on **mobilitics and justice** in examining the impacts of Covid-19 bordering regulations in seafarers' worlds and how they shaped daily experience. Finally, in "Infrastructural Containment and the Politics of Migration in the Mediterranean Sea", Lara Şarlak's angle is to interrogate the ocean as a space of **control and conflict** in unpacking the politics of migration in the Mediterranean Sea and the role of infrastructures of containment in bordering the movements of people.

Reading Across Borders

Throughout these chapters, the book aims to offer as much to scholars of the seas and of ocean governance, to those concerned with borders per se (following Turner and Peters' assertion in the Call for Papers for the conference session framing this book that the sea becomes a useful frame for thinking about border politics at large). As such, we suggest the book transcends borders—it is both a book *of* the ocean but *beyond* it. We hope it gives border scholars some interesting points of reflection and finds resonance with those who are interested in topics as wide as labour rights, migration, conservation management, resource planning, tourism geographies, and beyond—all themes covered in this collection.

In addition, the text brings together not just a diversity of disciplines (as previously mentioned, crossing borders here too) but of authors from across the globe, diverse in gender, race, ethnicity and age (and intersections of these). The book convenes exciting postgraduate talent, emerging postdoctoral scientists, lecturers as well as professors and those working beyond the academy in practice. Accordingly, the book aims to break some of the usual borders of scholarship—showcasing a range of scholars doing innovative work on borders but combined by their study of such processes at sea.

Finally, the book can be read as a whole—transgressing the borders of each chapter towards a combined narrative on the dynamics of ocean bordering. Indeed, those engaging the text are encouraged to voyage thematically and theoretically across the chapter boundaries to grasp some of the larger questions asked about the logics of bounding fluid space. Simultaneously, the contributors each also offer sealed contributions that can be dipped into and out of, as interest dictates. Yet together they do present a sustained attempt to grapple with how the 'freest' of spaces is enclosed: how borders are imagined, realised, planned, contested, experienced and lived at sea. Together, and separately, they also consider the possibilities that might arise if we think *beyond* borders as modes of oceanic management. We return to this discussion in "Conclusion: Openings—Ocean Governance (Beyond) Borders". For now, we invite readers to dive in, to examine with us and our contributors, the enduring, ongoing, and proliferating use of borders, to take seriously the strange insistence of line drawing in liquid spaces for methods of management and control (Lambach, 2021; Peters, 2020) and to question the very basis for bordering and their alternatives.

References

Adams, W. B. (2013). *Against extinction: The story of conservation.* Routledge

Adams, W. M. (2020). Geographies of conservation III: Nature's spaces. *Progress in Human Geography, 44*(4), 789–801.

Beery, J. (2016). Unearthing global natures: Outer space and scalar politics. *Political Geography, 55*, 92–101.

Bennett, N. J., Govan, H., & Satterfield, T. (2015). Ocean grabbing. *Marine Policy, 57*, 61–68.

Braverman, I. (2022). Amphibious legal geographies: Toward land–sea regimes. In I. Braverman (Ed.), *Laws of the sea: Interdisciplinary currents* (pp. 1–25). Routledge.

Campling, L., & Havice, E. (2014). The problem of property in industrial fisheries. *The Journal of Peasant Studies, 41*(5), 707–727.

Commission for the Convention on Antarctic Marine Living Resources (CCAMLR). (2013). *Progress report on the scientific data compilation and analyses in support of the development of a CCAMLR MPA in the Weddell Sea (Antarctica) by the Delegation of Germany.* Retrieved June 19, 2024, from https://epic.awi.de/id/eprint/34191/1/sc-xxxii-bg-07.pdf

Choi, Y. R. (2022). Slippery ontologies of tidal flats. *Environment and Planning E: Nature and Space, 5*(1), 340–361.

Cresswell, T. (1992). *In place/out of place: Geography*. The University of Wisconsin-Madison.

Cresswell, T. (2014). *Place: An introduction*. John Wiley & Sons.

Day J., Dudley, N., Hockings, M., Holmes, G., Laffoley, D., Stolton, S., Wells S., & Wenzel L. (Eds.) (2019). *Guidelines for applying the IUCN Protected Area Management Categories to Marine Protected Areas. Second Edition*. IUCN. Retrieved June 19, 2024, from https://portals.iucn.org/library/sites/library/files/documents/PAG-019-2nd%20ed.-En.pdf

Dickson, A. J. (2021). The carceral wet: Hollowing out rights for migrants in maritime geographies. *Political Geography, 90*, 102475. https://doi.org/10.1016/j.polgeo.2021.102475

Elden, S. (2010). Land, terrain, territory. *Progress in Human Geography, 34*(6), 799–817.

Fairbanks, L., Campbell, L. M., Boucquey, N., & St. Martin, K. (2018). Assembling enclosure: Reading marine spatial planning for alternatives. *Annals of the American Association of Geographers, 108*(1), 144–161.

Foley, P., & Mather, C. (2019). Ocean grabbing, terraqueous territoriality and social development. *Territory, Politics, Governance, 7*(3), 297–315.

George, R. Y., & Wiebe, S. M. (2020). Fluid decolonial futures: Water as a life, ocean citizenship and seascape relationality. *New Political Science, 42*(4), 498–520.

Gray, N. J. (2018). Charted waters? Tracking the production of conservation territories on the high seas. *International Social Science Journal, 68*(229–230), 257–272.

Hung, P. Y., & Lien, Y. H. (2022). Maritime borders: A reconsideration of state power and territorialities over the ocean. *Progress in Human Geography, 46*(3), 870–889.

Jay, S. (2010). Built at sea: Marine management and the construction of marine spatial planning. *The Town Planning Review, 81*(2), 173–191.

Jay, S. (2020). Measured as the water flows: The striated and smooth in marine spatial planning. *Maritime Studies, 19*(3), 255–268.

Johnson, C., Jones, R., Paasi, A., Amoore, L., Mountz, A., Salter, M., & Rumford, C. (2011). Interventions on rethinking 'the border' in border studies. *Political Geography, 30*(2), 61–69.

Johnston, R., Gregory, D., Pratt, G., & Watts, M. (Eds.). (2002). *The dictionary of human geography*. Blackwell.

Jonas, A., & While, A. (2005). Governance. In D. Atkinson, P. Jackson, D. Sibley, & N. Washbourne (Eds.), *Cultural geography: A critical dictionary of key concepts* (pp. 72–79). IB Tauris.

Jones, H. (2016). Lines in the ocean: Thinking with the sea about territory and international law. *London Review of International Law, 4*(2), 307–343.

Jones, O. (2011). Lunar–solar rhythm patterns: Towards the material cultures of tides. *Environment and Planning A, 43*(10), 2285–2303.

Jones, R. (2009). Categories, borders and boundaries. *Progress in Human Geography, 33*(2), 174–189.

Jones, R. (2016). *Violent borders: Refugees and the right to move*. Verso Books.

Khalili, L. (2021). *Sinews of war and trade: Shipping and capitalism in the Arabian Peninsula*. Verso Books.

Kothari, U. (2020). Between the land and the sea: Refugee experiences of the lighthouse as a real and symbolic border. *Borderlands, 19*(1), 63–88.

Lambach, D. (2021). The functional territorialization of the high seas. *Marine Policy, 130*, 104579. https://doi.org/10.1016/j.marpol.2021.104579

Langewiesche, W. (2010). *The Outlaw Sea: A world of freedom, chaos, and crime*. North Point Press.

Loyd, J. M., Mitchell-Eaton, E., & Mountz, A. (2016). The militarization of islands and migration: Tracing human mobility through US bases in the Caribbean and the Pacific. *Political Geography, 53*, 65–75.

Maxwell, S. M., Hazen, E. L., Lewison, R. L., Dunn, D. C., Bailey, H., Bograd, S. J., & Benson, S. (2015). Dynamic ocean management: Defining and conceptualizing real-time management of the ocean. *Marine Policy, 58*, 42–50.

McCorristine, S., & Adams, W. M. (2020). Ghost species: Spectral geographies of biodiversity conservation. *Cultural Geographies, 27*(1), 101–115.

Moran, D., Turner, J., & Schliehe, A. K. (2018). Conceptualizing the carceral in carceral geography. *Progress in Human Geography, 42*(5), 666–686.

Mountz, A. (2020). *The death of asylum: Hidden geographies of the enforcement archipelago*. University of Minnesota Press.

Mountz, A., & Hiemstra, N. (2012). Spatial strategies for rebordering human migration at sea. In T. M. Wilson & D. Hastings (Eds.), *A companion to border studies* (pp. 455–472). Wiley.

Newman, D. (2006). The lines that continue to separate us: Borders in our borderless world. *Progress in Human Geography, 30*(2), 143–161.

Nyman, E. (2019). Techno-optimism and ocean governance: New trends in maritime monitoring. *Marine Policy, 99*, 30–33.

Paasi, A. (2012). Border studies reanimated: Going beyond the territorial/relational divide. *Environment and Planning A, 44*(10), 2303–2309.

Paasi, A. (2013). Borders and border crossings. In N. C. Johnston, R. H. Schein, & J. Winders (Eds.), *The Wiley-Blackwell companion to cultural geography* (pp. 478–493). John Wiley & Sons.

Paasi, A. (2016). Borders. In K. Dodds & M. Kuus (Eds.), *The Ashgate research companion to critical geopolitics* (pp. 213–229). Routledge.

Peters, K. (2014). Tracking (im)mobilities at sea: Ships, boats and surveillance strategies. *Mobilities, 9*(3), 414–431.

Peters, K. (2020). The territories of governance: Unpacking the ontologies and geophilosophies of fixed to flexible ocean management, and beyond. *Philosophical Transactions of the Royal Society B, 375*(1814), 20190458. https://doi.org/10.1098/rstb.2019.0458

Peters, K., & Steinberg, P. E. (2019). The ocean in excess: Towards a more-than-wet ontology. *Dialogues in Human Geography, 9*(3), 293–307.

Peters, K., Steinberg, P. E., & Stratford, E. (Eds.) (2018). *Territory beyond terra.* Rowman & Littlefield.

Ryan, A. (2012). *Where land meets sea: Coastal explorations of landscape, representation and spatial experience*. Ashgate.

Steinberg, P. E. (1999). Lines of division, lines of connection: Stewardship in the world ocean. *Geographical Review, 89*(2), 254–264.

Steinberg, P. E. (2001). *The social construction of the ocean.* Cambridge University Press.

Steinberg, P. E., & Peters, K. (2015). Wet ontologies, fluid spaces: Giving depth to volume through oceanic thinking. *Environment and Planning D: Society and Space, 33*(2), 247–264.

Stierl, M. (2021). The Mediterranean as a carceral seascape. *Political Geography, 88*, 102417. https://doi.org/10.1016/j.polgeo.2021.102417

Teschke, K., Brtnik, P., Hain, S., Herata, H., Liebschner, A., Pehlke, H., & Brey, T. (2021). Planning marine protected areas under the CCAMLR regime–The case of the Weddell Sea (Antarctica). *Marine Policy, 124*, 104370. https://doi.org/10.1016/j.marpol.2020.104370

Turner, J. (2016). *The prison boundary: Between society and carceral space.* Palgrave Macmillan.

Urbina, I. (2019). *The outlaw ocean: Journeys across the last untamed frontier*. Vintage.

Vadrot, A. M. (2020). Ocean protection. In J.-F. Morin & A. Orsini. (Eds.), *Essential concepts of global environmental governance* (pp. 173–174). Earthscan Routledge.

BY

Overdetermined by Territory? Governing the Ocean in Time, Matter, and Rhythm

Christopher McAteer

Introduction

The global ocean can at times seem impervious to coherent management (see Acton et al., 2019). The pushback of migrant boats in the Mediterranean by EU and national agencies violates human rights principles (Barnes, 2022; Stierl, 2022 and Şarlak, this volume); illegal overfishing goes relatively unchecked by weak governance regimes that can also criminalise independent, particularly Indigenous, fishing (Long et al., 2020; Satizábal et al., 2021); and rapid Arctic sea ice loss appears to open myriad geopolitical problems that potentially threaten the very basis of international cooperation (Steinberg et al., 2015). As climate breakdown continues to put even greater stress on the health of the ocean and threaten low-lying areas of human habitation, it is more pressing than ever to ask why ocean governance so often seems to fail and how it can

C. McAteer (✉)
York University, Toronto, ON, Canada
e-mail: c.t.mcateer@gmail.com

K. Peters and J. Turner (eds.), *Ocean Governance (Beyond) Borders*,
https://doi.org/10.1007/978-3-031-71322-4_2

be done better. This is even more vital at a time in which there is a proliferation of ocean governance emerging, such as the Biodiversity Beyond National Jurisdiction (BBNJ) Treaty, that opens new frontiers in the protection of marine biodiversity and the high seas, but the significance and meaningfulness of which is yet to be understood.

In this chapter, I ask whether a major issue in ocean governance may be an ontological one (see also Conde et al., 2022; Peters, 2020; Steinberg & Peters, 2015). Specifically, I ask whether existing ocean governance mechanisms too often take the spatial character of the ocean for granted, which ultimately act as forms of territorialisation and bordering by folding ocean space into the logic of national terrestrial space via what Elden would term the 'technology of territory' (Elden, 2013). I therefore argue that to better theorise the ocean, it is necessary to contend not just with the abstract notion of space but with temporality and materiality too, which requires a spatial ontology that is formed around analytics of process and complexity, rather than boundaries and containers. But such an ontological approach would seem to be in friction with the dominant forms of ocean governance that exist today even as it is in line with the way many ocean theorists think (Lambach, 2021; Peters, 2020). Following Maxwell et al., I ask how ocean governance can be envisaged without a coherent figure of the ocean, focusing our attention instead on the fluid, processual, and chaotic nature of ocean space; the patterns of movement and rhythms of matter in time (Maxwell et al., 2015).

In the chapter that follows, I use a variety of examples from the '30 × 30' campaign that seeks to create marine protection for 30% of the global ocean by 2030 (see Peters, 2021; Royal Museums Greenwich, n.d.) to the governance of jellyfish blooms (Rothe, 2020) to develop a broader philosophical point on how we understand the oceans, and how those understandings shape governance. In doing so, I trace several key scholarly works that have moved geographical thinking beyond a planar and container conceptualisation of the ocean and towards a space of three and even four dimensions (see Childs, 2020). Such a tracing is productive for rethinking how ocean governance is designed in the first place, opening up new ways of thinking about ocean space that escape the sovereign technology of territory and, what Agnew would term, the 'territorial trap' in which borders and other boundaries become reified and monolithic and are assumed to be integral to managing the environment. Instead, I pay attention to how the volume and depth of space affects how the ocean is governed, as well as the necessity of thinking about matter and process as being an integral part of space. In doing so, I develop in this chapter a spatial ontology of the ocean that accounts for volume, fluid matter, and, importantly, rhythm (i.e., time). In conclusion, I argue that theorising the materiality and temporality of the ocean raises a particular set of problems for ocean governance due to the ontological shift that is required to grasp the complexities of ocean space and how this rubs up against the 'terrestrial limits' that undergird dominant forms of governance.

To progress this line of argument, it is necessary to first explore some of the tensions between dominant forms of governance and ocean space itself. In other words, it is vital to ask how ocean space is sometimes being conceptualised poorly and how such conceptualisations are shaping forms of governance and directing what is being governed. This now follows in two sections: 'Governing the surface' and 'From surface to container'. I then use these sections to open space for conceptualising differently—beyond these borders of thinking—the oceans and their governance. Here I consider the importance of matter, time and, linked accordingly, rhythm and relationality.

Governing the Surface

In June of 2023, the UN adopted the BBNJ Treaty, a legally binding treaty for the governance of international waters (UN, 2023). It came in the wake of the Global Biodiversity Framework that set a target of protecting 30% of the planet's land and seas by 2030, an initiative that had been keenly campaigned on by environmental groups pushing for greater protection of the ocean (Royal Museums Greenwich, n.d..; UNCBD, 2021). The need for such protection, the argument goes, could hardly be greater. Pollution, overfishing, warming waters, and acidification are destabilising the ocean to the point that 'we are entering an unknown territory of marine ecosystem change' (Bijma et al., 2013: 495) in which both marine and human health are at great risk (Dermawan et al., 2022). At the time of writing, only 8.2% of the ocean is enclosed within marine protected areas (MPAs)—one of the most common forms of marine protection—and only 2.9% is considered highly protected (Marine Conservation Institute, n.d.). Furthermore, these statistics vary depending on the sources consulted (e.g., see Protected Planet, n.d.) due to the lack of coherency in how MPAs are designated. The 30 × 30 campaign has argued that expanding the number, size, and protection level of MPAs can effectively safeguard and restore the biodiversity of the ocean and therefore the associated ecosystem services that it provides, such as being a crucial planetary carbon sink (Sala et al., 2021). However, some scholars have recently raised questions as to whether such an increase in the number of MPAs can actually be secured and respected by states and industries, and indeed whether tools such as MPAs can sufficiently deliver for marine biodiversity in the first place (Eisen & Mudodosi, 2021; Leenhardt et al., 2013; Pendleton et al., 2018). These questions about the effectiveness of MPAs and other forms of marine spatial planning (MSP) are certainly valid ones, but, I argue, there are other, more pertinent questions that demand answers first. These centre on what ocean space actually is, how it differs from terrestrial space, and what is actually being governed (see Acton et al., 2019; Boucquey et al., 2016; Conde et al., 2022; Maxwell et al., 2015; Peters, 2020; Steinberg & Peters, 2015).

By adopting a language of percentages and areas, the 30 × 30 campaign (and MSP approaches more generally) arguably relies on the notion that the ocean can be meaningfully measured within two dimensions, regardless of its depth (see Peters, 2021). This raises deep ontological questions about how ocean space is being conceived of within dominant governance regimes. The drawing of lines onto flat cartographic representations of ocean space speaks exclusively to the surface of the ocean (Peters, 2021) and reveals a reliance on what Elden terms the 'technology of territory', a '[technique] for measuring land and controlling terrain' that is fundamentally political in nature and reliant on the technology of bordering (Elden, 2013: 36). In this way, such an approach to space that foregrounds percentages, zones, and areas may come to overdetermine the ocean as a territorial space first and foremost, relegating specifically marine ecological concerns due to the undergirding logic of politics. This should perhaps not be entirely surprising when we consider that the management of oceans has historically pivoted on the concerns of capital and state, taking the form of access to resource extraction and the securitisation of territorial waters (Campling & Colás, 2018; Steinberg, 2001). In other words, the way that 30 × 30 conceptualises ocean space seems to rely upon the same ontology of space that subtends sovereignty, which is problematic because marine ecological issues stemming from pollution, overfishing, warming waters, and ocean acidification are not neatly bound up within clearly demarcated boundaries.

The problem is that the 'stuff' that is being governed in ocean governance does not stay within sovereign borders and other such politically determined boundaries. This problem is not, of course, unique to the ocean and also applies to dry land, the sky, and the spaces between, as Peters et al. (2018) have demonstrated, among others (see also Zee, 2020). However, this is further complicated at sea by the fact that, as Peters has noted elsewhere, the ocean is 'geophysically liquid, mobile, three-dimensional' (Peters, 2020: 6), subject to constant process and change, and much of what is being protected in enclosures such as MPAs is similarly mobile (see Maxwell et al., 2015). Fixed lines on a map that demarcate protected areas therefore tend to fall into what John Agnew refers to as 'the territorial trap', that is, a way of theorising space that is bounded to geophysical features that appear to be fixed and solid but which are actually flexible and regularly altered (Agnew, 1994). Therefore, if dominant modes of ocean governance are falling into this trap at least in part due to the way they conceptualise space, specifically by using a language of space that seems more suited to the terrestrial (see Steinberg & Peters, 2015), it may be necessary to begin a study of ocean governance by attending to some important ontological questions about ocean space first.

This is the challenge that Peters lays down in a recent critique of established and nascent ocean management approaches (Peters, 2020). Noting the relatively slight engagement with questions of ontology in the practice and scholarship of ocean governance, Peters seeks to unsettle overly land-based classifications and explanations of what the ocean is. Unlike epistemology, the study of knowledge or how we know something, ontology opens up the question of what we know in the first place. In other words, ontology is about the conditions of something's existence, the nature and form that a thing takes in the world, how it exists as a being distinct from our perception of it. As Peters argues, ontological questions often get buried within the practice and study of ocean governance while the epistemological questions get advanced further. This lack of sufficient engagement in ontological questions creates a problem in which scholars and practitioners of ocean governance too often fail to ask big questions about what the ocean is, instead focusing on how we can access knowledge about it (see Lambach, 2021). In other words, the nature of the ocean is often assumed as determined and the questions tend towards how it can be measured and managed (see Sammler, 2020). This may in part account for the tendency to transplant spatial approaches that rely on percentages, areas, and zones to the governing of the ocean without always questioning whether forms of governance that were developed on and for dry land can really be applied so easily to the ocean. These arguments are made especially well in the literature on MSP, which has its roots very firmly in grounded modes of planning (Jay, 2018, 2020). We therefore need to complicate and reassess the ontological assumptions that undergird dominant bordered approaches to ocean governance before we can begin to speculate on new forms of governance that avoid inscribing the ontological assumptions of territory and the logic of politics into projects of marine conservation, and thus onto conceptions of ocean space itself.

From Surface to Container

Over the past decade or so, geographers have powerfully critiqued areal approaches to analysing space and in doing so have complicated the ontological grounding of space itself. One of the ways this has been done is by engaging with all three dimensions of space. This has resulted something of a 'volumetric turn' in geography and related disciplines

(Jackman & Squire, 2021), a theoretical shift that has quite literally deepened scholarly thinking on space and territory. Scholarship has ranged from retheorising the spatial axes on which power operates (Elden, 2013; Weizman, 2002), to reconceptualising the volume of aerial (Adey, 2010; Jackman, 2017; Sloterdijk, 2009), subterranean (Garrett, 2016; Slesinger, 2020; Squire & Dodds, 2020), and marine spaces (Peters & Steinberg, 2019; Squire, 2017; Steinberg & Peters, 2015). Elden, in particular (2013), opened up the question of volume in an influential paper about the dimensions of territory and security. In exploring the problems that arise on dry land when territory is understood simply as an area, Elden implored geographers to 'think above and below, to conceptualise space in three dimensions' rather than just two (2013: 49). The point for Elden is that power operates to produce and reproduce territory in three dimensions, up and down as well as along a plane, in bounding and enclosing sovereign space. To focus only on two dimensions therefore misses out on the ways in which states territorialise the space above and below the ground. We only need to consider the means by which the air is demarcated and secured by states through the monitoring and policing of national airspace (Williams, 2010; see also Lin, 2016), or for example, how the Iraqi government in 1990 claimed that Kuwait was using 'slant drilling' techniques to steal oil across its border, thereby violating its subterranean sovereignty (Meierding, 2022). But, as Elden argues, this does not mean that three-dimensional territory is a settled space of control. Rather, it is a process that is constantly evolving and in flux. Through a consideration of the spatial politics of the West Bank and Gaza, building on Weizman, Elden suggests an innovative reconceptualisation of geo-politics that mimics the 'calculation and metrics' of biopolitics. In other words, geo-politics is a spatial form of governmentality. This approach opens new ways of thinking through what space is by theorising how power operates within it in complex and multivalent ways in all three dimensions, similar to Ingold's understanding of space as entangled (2008) or Sloterdijk's spherical thinking (2011).

Steinberg and Peters (2015) innovatively took key aspects of Elden's theoretical intervention on volumetric space and cast it out to sea. Arguing that geographical thinking needs to study the world beyond 'the plane geometry of points, lines, and areas that have long grounded the discipline', they note that while the legal control of the sea has been discussed at length by scholars of various disciplines, few have explored its space as volume (2015: 248). In the context of ocean governance, a turn to volume would seem to offer a better approach to determining the space that is being governed, accounting for depth as well as surface area. However, as Elden might remind us (2013), this would not escape an alignment to the logic of sovereignty and spatial forms of governmentality that push ecological concerns to the background. Also, as Steinberg and Peters argue (2015), while thinking through volume helps us to resist the flattening of space, if we wish to comprehend the ocean more fully, we must think beyond volumetric space as an empty vessel. In other words, while bringing volume to the ocean moves beyond the two-dimensional planar approach by adding the vertical, thinking of the ocean as simply a container fundamentally misunderstands that it is composed of fluid matter. Steinberg and Peters thus argue that we should foreground the processual and flowing nature of matter in ocean space, avoiding ideas of it as fixed and static. This is not to say that terrestrial space is fixed and static in its geophysical characteristics, but, as Maxwell et al. argue, 'the scale of temporal and spatial variability in the oceans is unmatched in terrestrial systems', suggesting that spatial management techniques derived from dry land may simply not be suited to the dynamism of the ocean (2015: 43).

This introduces yet another dimension to the spatial ontology of the ocean, that of time, because to think of process one must think of rhythm which requires an attendance to the temporal dimension of the ocean. Therefore, if the challenge is to develop a sounder ontological understanding of ocean space in order to design more appropriate and effective forms of marine governance that can better protect the ocean, then this means theorising both the material and temporal dimensions of ocean space together. To do this, it is necessary to think beyond the ocean as both surface *and* an empty container of space. If indeed it can be demonstrated that ocean space is complex, relational, and three or even four dimensional, then it is possible to explore how novel forms of governance can work with ocean space *as* ocean space, rather than ocean space as specifically political, territorial, and oftentimes bordered space. First, it is necessary to understand ocean space beyond its dimensions by engaging with the question of how its space is made. To do so, we must look to its materiality and the temporal processes within which its matter exists.

Beyond the Container

Matter

Steinberg and Peters argue that if we are to understand the ocean and human interactions with it, we need to attempt to 'think with the ocean', which requires a different spatial ontology from what is commonly deployed on dry land or even at the surface (2015: 248). They therefore draw on Doreen Massey's work that challenges Euclidean notions of space as stable and remote from the substances and matter that fill it, arguing along the lines of Massey that such conceptions of space make it impossible to understand immanent geographies and how they relate. In other words, if space is imagined as a container and the stuff within that container is inherently separate from it, then we are left with an entirely abstract conception of space that is constituted merely by boundaries and limits within a void. Seeking an alternative ontology of space that dispenses with the container notion and conceptualises matter as an inherent part of a given space, Steinberg and Peters draw on Marston et al.'s idea of 'flat ontology' (2005), an approach which flattens scalar hierarchies that grant particular nodes or objects greater precedence over others within a given space. Dispensing with scale, flat ontology instead focuses on sites or 'event-spaces' (Ash, 2020; Jones et al., 2007). This may point us to a theory of space that aligns closely with the literature of new materialism, particularly authors such as Bennett (2010a, 2010b) who argues that material things such as plastic waste, food, electricity networks, etc. have political agency, a 'thing-power', and that this 'vibrant matter' exists within a complex web of actors in the world that includes humans but is not overdetermined by them. However, Steinberg and Peters suggest that while the spatial ontology that emerges from such an approach makes it possible to conceptualise matter, such a conception 'fails to take account for the chaotic but *rhythmic* turbulence of the material world' (2015: 248). They note that bringing flat ontology into conversation with volumetric approaches can be helpful here, but that such an intervention can risk creating yet more abstract space, albeit much more complex an abstraction than the Euclidean model. Therefore, if we are to bring matter into this debate, we need to foreground the *processual and flowing nature of that matter* (be it plankton, fish, plastics, or whatever), avoiding the idea that it is fixed or static. To be clear, the question here is about ocean space on an ontological level, less about the specific matter that composes the ocean, be this at the molecular level of H_2O, etc. or the level of discernible objects such as boats and buoys or various marine species. While the 'stuff' that constitutes the material character of the ocean is indeed important, my focus here is more on how this stuff in general is implicated in the proliferating and complex relations that come to challenge ideas of ocean space as monolithic and container like.

Here, 'wet ontology' has resonance (Steinberg & Peters, 2015). The ocean is conceptualised not as 'a space of discrete points between which objects move, but rather as a dynamic environment of flows and continual recomposition where, because there is no static background, "place" can be understood only in the context of mobility' (2015: 257). This highly mobile and shifting conception of space demands that we consider another dimension to the ocean—that of time—because to think of flows, mobility, and circulation, one must think of rhythm and therefore the role that time plays in constituting ocean space. The challenge for the governance of ocean space can then be seen as one of how such a complex and relational space can be managed at all, a question that Maxwell et al. ask in their enquiry into how ocean governance can be 'as fluid in space and time as the resources and users we aim to manage' (2015: 43). In other words, any novel form of governance that seeks to comprehend ocean space *as* ocean space, rather than as sovereign or territorial space, must contend not only with the relational nature of matter in the ocean, but also with the question of *time*, because matter can only be understood as relational if it is understood as in process and thus in time.

Time

Over the past decade or so, there has been renewed attention in geography and international relations to the question of how time and temporality relate to territory and space (Ho, 2021; Rao, 2019). Indeed, both disciplines have a long history of engaging with questions about time and temporality in a variety of ways (e.g., Agathangelou & Killian, 2016; Hägerstrand, 1983; Harvey, 1989; Hom, 2020; Hutchings, 2008; Massey, 1992; May & Thrift, 2001; Nixon, 2011). Massey (1992, 1995, 2005) notably argued that the analysis of both the spatial and temporal was essential to social theory, rejecting tendencies existing in dominant social and political theories that attempt to formulate a theory of space that lacks an analysis of time. As Massey argued, 'territory is integrally spatio-temporal' and no place can exist in the world as a 'static instantaneity' (2005: 28). Therefore, in seeking a spatial ontology that can form the basis of ocean governance, we must contend with the fact that territory does not only exist within three dimensions, but rather four (following Childs, 2020). Drawing on Massey's arguments about the contingency of time, as well as Steinberg and Peters' attention to the circulation and rhythm of ocean space, I argue that, to theorise the space of the ocean more completely, we must examine temporality and volume together, which requires a spatial ontology that is formed around process and change, as well as matter and containers. Such an ontology of space would assert that space has an inherent temporal dimension, which would mean theorising change, process, and vacillation as more than a variable within a space, but rather an integral aspect of space itself.

Oceans Beyond Borders?

In a recent essay that theorises the spatial peculiarities of bordering practices in the Himalayas, Harris brings time into conversation with volume (2020). Exploring the difficulty that states have in controlling borders at high altitudes, Harris argues that not only are two dimensions insufficient for understanding the topography of territorial space, but so are the three of volumetric approaches, because 'there is also a teleological aspect to bordering practices' (2020: 86). Securing, controlling, and governing high altitude borders—particularly disputed ones such as the western Himalayan border between Pakistan, India, and China—is difficult to achieve on an infrastructural and practical level. The time it takes to send border guards to stations, or the constrictions on how long guards can remain on post due to weather conditions, renders such borders as both spatially *and* temporally insecure, revealing for Harris the temporal nature of bordering and sovereignty. This is similar to the insights of the 'Italian Limes' research project that studied the effects of climate change on the Italian-Austrian border along the Alpine range (Italian Limes, n.d.). The border, which is clearly marked along rocky parts of terrain, cannot be so easily demarcated at higher altitudes and so relies on the physical measurement of the landscape. However, with such considerable geophysical changes in the region due to glacier melt driven by global heating, the project revealed that, over time, the border has moved hundreds of metres in some areas. As a result, regular surveys had to be agreed between Italy and Austria to ascertain whether their borders needed to be redrawn on the official maps. The project thus highlighted how slippery and contingent the border is over time. In other words, if we are to understand the contingencies of territories, then we have to look beyond space as a static container—whether that be in two or three dimensions—and bring in the dimension of time too.

If we return our focus to ocean space, this is precisely where Steinberg and Peters' intervention is pushing us. That is, beyond the notion of the ocean as simply a container and towards an understanding of the ocean as a space of ongoing geography that is continually composed and recomposed in its spatial dimensions. This composition and recomposition occurs within time, not simply in a durational sense of time passing, but rather on a number of different temporal scales and directions all at once. Slow linear temporal processes such as the lowering of pH levels as CO_2 is absorbed into the ocean, or the gradual accumulation of debris in the 'Great Pacific garbage patch', exist alongside faster linear temporal events such as waves and passing weather. But non-linear temporal processes also exist alongside these linear ones, such as the circulatory and cyclical time of tides, seasonal sea ice, animal migrations, and the mass churning of molecules that never ceases within the ocean. The space of the ocean, once understood in depth, matter, *and* time, is revealed to be a space that is under constant change. Yet, states and international institutions maintain stubborn attempts to demarcate, enclose, and territorialise it—to enact and deploy borders—most commonly by conceiving the ocean as an abstract two-dimensional space. This insight perhaps aligns with Walker's argument (1993: 180) that the state normalises its sovereignty through the drawing of insides and outsides, through practices of inclusion and exclusion, that cut through the 'ontological density of the principle of state sovereignty'. It does this not only spatially, but specifically by reifying the eternality of the state while side-stepping deeper questions about time which might otherwise threaten the timeless epistemes of sovereignty (Walker, 1993: 180). Read in the context of oceans, bordering has become timeless or 'fixed' as an approach of governing (Lambach, 2021; Peters, 2020), yet greater attention to time, and notably rhythm and relationality, is necessary to avoid side-stepping such questions regarding the governance of the ocean.

Rhythm and Relationality

An attention to the fluid and churning nature of the ocean means it becomes impossible to think of the ocean as an object that can be divided up into discernible parts (such as the 30 × 30 aims to do) or studied as a singular whole (such as the 'one ocean' concept). Rather, we seem to require the sort of 'bottom-up' approach of assemblage theory, in which process is privileged above object; fluidity and fungibility privileged above the static and fixed (see DeLanda, 2006; Deleuze & Guattari, 2004). This approach forces a shift from isolated categories to networks composed of complex and multiple relations, which can be a helpful way to analyse governance regimes because the focus moves from distinct actors and actions to experimental processes and the unexpected emergence of relationships within a spatially complex and changing environment (Anderson, 2012). However, as Anderson has observed (2012), an assemblage approach can, when applied to space, lead to what he terms as a 'territorialised assemblage'. While the relationships between categories remain at the fore, the problem here is that the categories can become reified, their fungibility made impossible.

Accordingly, Anderson instead uses assemblage theory to de-territorialise ocean space, analysing the phenomenon of the surfed wave as a 'relational place', that is, like all places, 'not only geographical but temporal' (2012: 583). Anderson is thinking about time in a deep way here, arguing that 'the trajectories that come together to form relational places, and then leave to form others, operate at different velocities' (2012: 583). So, both time and space must be conceptualised in complex and multivalent ways. He seems to argue that place is produced by a convergence of both linear and non-linear temporalities, as well as geographically specific and relational spaces. This is helpful to understanding the dynamic nature of ocean space and how we might try to map and govern it differently.

Indeed, moving away from the idea of the ocean as a container of space to that of multiple, coexisting relational places can help us to de-territorialise and shift away from a sovereign ontology that is predicated upon territory and its oppressive operations of power that push ecological concerns aside. The scholarship mentioned above moves us towards a more entangled understanding of space that foregrounds relationality and complex convergences of multiple and sometimes unknowable beings. If we accept this ontological shift, then any traditional form of areal management begins to look untenable. But the question remains as to how we actually create and maintain governance that attends to a relationally oriented ontology of ocean space.

Here, I turn to a recent paper by Rothe (2020) that asks how technology can mediate entangled relations with the ocean. Rothe focuses on a novel form of management that seeks to cull jellyfish blooms that threaten marine ecosystems, critical infrastructure, aquaculture, and military vessels (Rothe, 2020). He reveals that the growth in large conflagrations of jellyfish in the ocean is becoming a serious security and economic threat, as well as harming biodiversity. This phenomenon is of interest to Rothe because it is not only impacting human and other animal relations with the oceans but is also in part a product of anthropogenic changes happening to the ocean, including warming, overfishing, and acidification, which are all amenable to jellyfish population growth. Rothe examines an experimental management program called the Jellyfish Elimination Robotic Swarm (JEROS), a swarm of autonomous swimming killer robots that is being deployed to detect and cull jellyfish. He considers the phenomenon of jellyfish blooms and how they are being dealt with through a relational approach that spreads agency among a variety of actors including humans, non-human animals, ecosystems, ships, nets, robots, and artificial intelligence (AI). While clearly indebted to violent technologies of the state, the JEROS program seems to exemplify a relational approach and, in many ways, the deployment of specifically self-learning AI with (albeit limited) free movement may in certain respects de-centre both sovereign territory and the human in interesting ways. Rothe argues that the use of autonomous robots in the mediation of ocean relations suggests a move away from a narrowly conceived top-down approach to governance. Rational management and control seem to be eschewed in favour of 'new forms of governance that work through complexity', seeking instead 'more experimental approaches' that can 'adapt to [sudden] environmental change' (Rothe, 2020: 147). Rothe is drawing here on the work of Chandler (2018) and the 'ontopolitics of the Anthropocene', which Rothe says can describe new forms of governance, 'that would give up the attempt to know, anticipate or control problems, and, instead, try to adapt to them or rework them towards more desirable ends' (Rothe, 2020: 147). This is referred to as 'ontopolitics' because the new and experimental forms of governance that emerge would work not on the epistemological level, but rather the ontological.

The aim, then, is not simply to gather and organise data and other abstract forms of knowledge for analysis by humans, but instead 'try to sense, modulate, or transform complex phenomena in the process of their emergence' (Rothe, 2020: 148). For Chandler, these experimental forms of governance are categorised as 'mapping, sensing, and hacking', each of which seeks to adapt to complex life-processes rather than control them. It is through an approach of entanglement that Rothe is thinking through the novel governance that Chandler proposes. This is about, as Rothe notes, governance that 'works through a logic of *sympoiesis* – or becoming with', (2020: 153) in order to create mechanisms and methods that do not overdetermine phenomena from an anthropocentric perspective nor further territorialise the ocean.

Conclusion

Bringing attention to matter and time, and to the four dimensions of ocean space (see Childs, 2020)—in sum asking ontological questions—may open up new theoretical pathways for ocean governance studies, but it remains methodologically ambiguous. Even if, as Steinberg and Peters (2015) compel us to do, we can develop governance regimes that 'think with the ocean' how can we ensure that they will be upheld from the most rapacious tendencies of capital and state sovereignty? This chapter does not contend that campaigns such as 30 × 30 or the emerging frameworks of the BBNJ Treaty are doomed to complete failure and should therefore be abandoned—they will likely provide some much-needed protection for marine ecosystems—but rather that they run the risk of reinscribing the lines and boundaries of territory onto a space that not only seems to resist territorialisation, but is also specifically being harmed by the technologies of sovereignty.

In recognising these risks, I argue that experimental forms of governance such as JEROS can perhaps be further studied in all of their messy complexity in order to start theorising multiple forms of governance that reject the search for global coherence at the expense of recapitulating the overdetermining logic of the state. By rethinking how we conceptualise ocean space at the ontological level, we may end up with more questions than answers. But this may be precisely the direction that governance—particularly of the environmental and ecological variety—needs to go in: away from the neat, totalising, and hierarchised regimes of the state, international institutions, and capital, and towards something experimental, adaptive, and thoroughly *in time* with the spaces it seeks to govern.

References

Acton, L., Campbell, L. M., Cleary, J., Gray, N. J., & Halpin, P. N. (2019). What is the Sargasso Sea? The problem of fixing space in a fluid ocean. *Political Geography, 68*, 86–100.

Adey, P. (2010). *Aerial life: Spaces, mobilities, affects*. Wiley-Blackwell.

Agathangelou, A. M., & Killian, K. D. (2016). *Time, temporality and violence in international relations: (De)fatalizing the present*. Routledge.

Agnew, J. (1994). The territorial trap: The geographical assumptions of international relations theory. *Review of International Political Economy, 1*(1), 53–80.

Anderson, J. (2012). Relational places: The surfed wave as assemblage and convergence. *Environment and Planning d: Society and Space, 30*, 570–587.

Ash, J. (2020). Flat ontology and geography. *Dialogues in Human Geography, 10*(3), 345–361.

Barnes, J. (2022). Torturous journeys: Cruelty, international law, and pushbacks and pullbacks over the Mediterranean Sea. *Review of International Studies, 48*(3), 441–460.

Bennett, J. (2010a). *Vibrant matter: A political ecology of things*. Duke University Press.

Bennett, J. (2010b). Thing-Power. In B. Braun & S. J. Whatmore (Eds.), *Political matter: Technoscience, democracy, and public life* (pp. 35–62). University of Minnesota Press.

Bijma, J., Pörtner, H.-O., Yesson, C., & Rogers, A. D. (2013). Climate change and the oceans – What does the future hold? *Marine Pollution Bulletin, 74*, 495–505.

Boucquey, N., Fairbanks, L., Martin, K. S., Campbell, L. M., & McCay, B. (2016). The ontological politics of marine spatial planning: Assembling the ocean and shaping the capacities of 'Community' and 'Environment.' *Geoforum, 75*, 1–11.

Campling, L., & Colás, A. (2018). Capitalism and the sea: Sovereignty, territory and appropriation in the global ocean. *Environment and Planning D: Society and Space, 36*(4), 776–794.

Chandler, D. (2018). *Ontopolitics in the anthropocene: An introduction to mapping*. Routledge.

Childs, J. (2020). Extraction in four dimensions: Time, space and the emerging geo(-)politics of deep-sea mining. *Geopolitics, 25*(1), 189–213.

Conde, M., Mondré, A., Peters, K., & Steinberg, P. E. (2022). Mining questions of 'what' and 'who': Deepening discussions of the seabed for future policy and governance. *Maritime Studies, 21*(3), 327–338.
DeLanda, M. (2006). *A new philosophy of society: Assemblage theory and social complexity*. Continuum.
Deleuze, G., & Guattari, F. (2004). *A thousand plateaus: Capitalism and schizophrenia*. University of Minnesota Press.
Dermawan, D., Wang, Y.-F., You, S.-J., Jiang, J.-J., & Hsieh, Y.-K. (2022). Impact of climatic and non–climatic stressors on ocean life and human health: A review. *Science of the Total Environment, 821*, 153387. https://doi.org/10.1016/j.scitotenv.2022.153387
Eisen, J., & Mudodosi, B. (2021, November 1). *30x30 – A brave new dawn or a failure to protect people and nature?* Green Economy Coalition. Retrieved May 1, 2024, from https://www.greeneconomycoalition.org/news-and-resources/30x30-a-brave-new-dawn-or-a-failure-to-protect-people-and-nature
Elden, S. (2013). Secure the volume: Vertical geopolitics and the depth of power. *Political Geography, 34*, 35–51.
Garrett, B. L. (2016). Picturing urban subterranea: Embodied aesthetics of London's sewers. *Environment and Planning A: Economy and Space, 48*(10), 1948–1966.
Hägerstrand, T. (1983). In search for the sources of concepts. In A. Buttimer (Ed.), *The practice of geography* (pp. 238–256). Longman.
Harris, T. (2020). Lag: Four-dimensional bordering in the Himalayas. In F. Billé (Ed.), *Voluminous states: Sovereignty, materiality, and the territorial imagination* (pp. 78–90). Duke University Press.
Harvey, D. (1989). *The condition of postmodernity*. Blackwell.
Ho, E.L.-E. (2021). Social geography I: Time and temporality. *Progress in Human Geography, 45*(6), 1668–1677.
Hom, A. R. (2020). *International relations and the problem of time*. Oxford University Press.
Hutchings, K. (2008). *Time and world politics: Thinking the present*. Manchester University Press.
Ingold, T. (2008). Bindings against boundaries: Entanglements of life in an open world. *Environment and Planning A: Economy and Space, 40*(8), 1796–1810.
Italian Limes. (n.d.) *Moving borders: A cartographic and political enquiry*. Retrieved May 1, 2024, from http://www.italianlimes.net/project.html
Jackman, A. (2017, June 27). *Sensing, speaking volumes*. Cultural Anthropology. Retrieved May 1, 2024, from https://culanth.org/fieldsights/sensing

Jackman, A., & Squire, R. (2021). Forging volumetric methods. *Area, 53*, 492–500.

Jay, S. (2018). The shifting sea: From soft space to lively space. *Journal of Environmental Policy & Planning, 20*(4), 450–467.

Jay, S. (2020). Measured as the water flows: The striated and smooth in marine spatial planning. *Maritime Studies, 19*(3), 255–268.

Jones, J. P. III, Woodward, K., & Marston, S. A. (2007). Situating Flatness. *Transactions of the Institute of British Geographers. New Series, 32*(2), 264–276.

Lambach, D. (2021). The functional territorialization of the high seas. *Marine Policy, 130*, 104579.

Leenhardt, P., Cazalet, B., Salvat, B., Claudet, J., & Feral, F. (2013). The rise of large-scale marine protected areas: Conservation or geopolitics? *Ocean & Coastal Management, 85*, 112–118.

Lin, W. (2016). Drawing lines in the sky: The emotional labours of airspace production. *Environment and Planning A: Economy and Space, 48*(6), 1030–1046

Long, T., Widjaja, S., Wirajuda, H., & Juwana, S. (2020). Approaches to combatting illegal, unreported and unregulated fishing. *Nature Food, 1*, 389–391.

Marine Conservation Institute. (n.d.). *The marine protection atlas*. Marine Conservation Institute. Retrieved November 5, 2022, from https://mpatlas.org/

Massey, D. (1992). Politics of space/time. *New Left Review, 196*, 65–84.

Massey, D. (1995). Places and their pasts. *History Workshop Journal, 39*, 182–192.

Massey, D. (2005). *For space*. Sage.

Maxwell, S. M., Hazen, E. L., Lewison, R. L., Dunn, D. C., Bailey, H., Bogard, S. J., Briscoe, D. K., Fossette, S., Hobday, A. J., Bennet, M., Benson, S., Caldwell, M. R., Costa, D. P., Dewar, H., Eguchi, T., Hazen, L., Kohin, S., Sippel, T., & Crowder, L. B. (2015). Dynamic ocean management: Defining and conceptualizing real-time management of the ocean. *Marine Policy, 58*, 42–50.

May, J., & Thrift, N. (2001). *Timespace: Geographies of temporality*. Routledge.

Meierding, E. (2022). Oil, materiality, and interstate war. In R. Dannreuther & W. Ostrowski (Eds.), *Handbook on oil and international relations* (pp. 79–93). Edward Elgar Publishing.

Nixon, R. (2011). *Slow violence and the environmentalism of the poor*. Harvard University Press.

Pendleton, L. H., Ahmadia, G. N., Browman, H. I., Thurstan, R. H., Kaplan, D. M., & Bartolino, V. (2018). Debating the effectiveness of marine protected areas. *ICES Journal of Marine Science, 75*(3), 1156–1159.

Peters, K. (2020). The territories of governance: Unpacking the ontologies and geophilosophies of fixed to flexible ocean management, and beyond. *Philosophical Transactions of the l Society B, 375*, 20190458. https://doi.org/10.1098/rstb.2019.0458

Peters, K. (2021, September 3). A line in the ocean 30x30: Ocean borders and geography's limits [Chair's Plenary Lecture]. l Geographical Society (with IBG) Conference, London, UK.

Peters, K., & Steinberg, P. E.(2019). The ocean in excess: Towards a more-than-wet ontology. *Dialogues in Human Geography, 9*(3), 293–307. https://doi.org/10.1177/2043820619872886

Peters, K., Steinberg, P. E., & Stratford, E. (Eds.). (2018). *Territory beyond terra.* Rowman & Littlefield.

Protected Planet. (n.d.). *Marine protected areas.* Retrieved November 5, 2022, from https://www.protectedplanet.net/en/thematic-areas/marine-protected-areas

Rao, R. (2019). One time, many times. *Millennium: Journal of International Studies, 47*(2), 299–308.

Rothe, D. (2020). Jellyfish encounters: Science, technology and security in the Anthropocene ocean. *Critical Studies on Security, 8*(2), 145–159.

Royal Museums Greenwich. (n.d.). *What exactly is 30x30?* Royal Museums Greenwich. Retrieved November 6, 2022, from https://www.rmg.co.uk/stories/our-ocean-our-planet/what-is-30x30-marine-protected-areas-ocean-2030

Sala, E., Mayorga, J., Bradley, D., Reniel, B. C., Atwood, T. B., Auber, A., Cheung, W., Costello, C., Ferretti, F., Friedlander, A. M., Gaines, S. D., Garilao, C., Goodell, W., Halpern, B. S., Hinson, A., Kaschner, K., Lesner-Reyes, K., Leprieur, F., McGowan, J., … Lubchenco, J. (2021). Protecting the global ocean for biodiversity, food and climate. *Nature, 592*, 397–402.

Sammler, K. G. (2020). The rising politics of sea level: Demarcating territory in a vertically relative world. *Territory, Politics, Governance, 8*(5), 604–620.

Satizábal, P., Le Billon, P., Belhabib, D., Saavedra-Díaz, L. M., Figueroa, I., Noriega, G., & Bennett, N. J. (2021). Ethical considerations for research on small-scale fisheries and blue crimes. *Fish and Fisheries, 22*(6), 1160–1166.

Slesinger, I. (2020). A cartography of the unknowable: Technology, territory and subterranean agencies in Israel's management of the Gaza tunnels. *Geopolitics, 25*, 17–42.

Sloterdijk, P. (2009). Airquakes. (E. T. Medieta, trans.) *Environment and Planning D: Society and Space, 27*(1), 41–57.

Sloterdijk, P. (2011). *Bubbles: Spheres volume I: Microspherology*. MIT Press.

Squire, R. (2017). Do you dive?": Methodological considerations for engaging with "volume. *Geography Compass, 11*, e12319. https://doi.org/10.1111/gec3.12319

Squire, R., & Dodds, K. (2020). Introduction to the special issue: Subterranean. *Geopolitics, 25*, 4–16.

Steinberg, P. E. (2001). *The social construction of the ocean.* Cambridge University Press.

Steinberg, P. E., & Peters, K. (2015). Wet ontologies, fluid spaces: Giving depth to volume through oceanic thinking. *Environment and Planning D: Society and Space, 33*, 247–264.

Steinberg, P. E., Tasch, J., & Gerhardt, H., (with Keul, A., & Nyman, E. A.). (2015). *Contesting the Arctic: Politics and imaginaries in the circumpolar North*. I. B. Taurus.

Stierl, M. (2022). Do no harm? The impact of policy on migration scholarship. *Environment and Planning C: Politics and Space, 40*(5), 1083–1102.

United Nations (UN). (2023, June 19). *Statement at the Intergovernmental Conference on an international legally binding instrument under the United Nations Convention on the Law of the Sea on the Conservation and Sustainable Use of Marine Biological Diversity of Areas Beyond National Jurisdiction.* United Nations. Retrieved May 1, 2024, from https://www.un.org/bbnj/sites/www.un.org.bbnj/files/06-15-2023-final_bbnj_statement.pdf

United Nations Convention on Biological Diversity (NCBD). (2021, July 12). *A new global framework for managing nature through 2030: 1st detailed draft agreement debuts.* United Nations. Retrieved May 1, 2024, from https://www.un.org/sustainabledevelopment/blog/2021/07/a-new-global-framework-for-managing-nature-through-2030-1st-detailed-draft-agreement-debuts/

Walker, R. (1993). *Inside/outside: International relations as political theory.* Cambridge University Press.

Weiqiang, L. (2016). Drawing lines in the sky: The emotional labours of airspace production. *Environment and Planning A: Economy and Space, 48*(6), 1030–1046.

Weizman, E. (2002, April 23). *Introduction to the politics of verticality*. Open Democracy. Retrieved May 1, 2024, from https://www.opendemocracy.net/en/article_801jsp

Williams, A. J. (2010). Reconceptualising spaces of the air: Performing the multiple spatialities of UK military airspaces. *Transactions of the Institute of British Geographers, 36*(2), 253–267.

Zee, J. (2020). Downwind. In F. Billé (Ed.), *Voluminous states: Sovereignty, materiality, and the territorial imagination* (pp. 119–130). Duke University Press.

Counter-Mapping: A Morphology of Oscillating Margins in the Norwegian Sea

Nancy Couling

Examining Borders at Sea: Making an Urban Ocean

In Western Europe, the history of the (attempted) division of ocean space parallels the landed history of territorial appropriation as an expression of political power. At sea, alliances and rivalries between influential individuals or political units were fluid and dynamic, oscillating between efforts to create a *Mare Clausum*—a closed sea—and a *Mare Liberum* (Grotius, 1609). Between the tenth and thirteenth centuries, the Norwegian Kings attempted to instate the former (Theutenberg, 1984), as did King James I of England in the seventeenth century, culminating in the famous 1635 publication *Mare Clausum* by John Selden on his behalf (Selden, 1663). Selden's *Mare Clausum* publication was a reaction to the preceding *Mare Liberum*, which famously declared the seas as belonging to all mankind (sic)—a seemingly ethical argument which conveniently served Dutch

N. Couling (✉)
Bergen School of Architecture, Bergen, Norway
e-mail: nancycouling@bas.org

K. Peters and J. Turner (eds.), *Ocean Governance (Beyond) Borders*,
https://doi.org/10.1007/978-3-031-71322-4_3

trading interests at the time and was intended to challenge the attempts of their English rivals to claim parts of the valuable sea space and its trading corridors.

Such abstract lines at sea were periodically materialised through military action. Marine blockades attempted to disrupt enemy supplies, as in the 1714 Battle of Gangut in the Baltic Sea where the Russian fleet succeeded in breaking through the Swedish blockade, resulting in the first naval victory for Czar Peter the First and the reopening of the Russian maritime supply lines. Such lines were temporary, incomplete, contested, frequently penetrated, and not yet solidified into borders.

The ocean resisted enclosure. Its spatial fluidity continued to be upheld through the *Mare Liberum* concept right until the UN Convention on the Law of the Sea in 1982 (UN, 1982). Prompted by coastal nations increasing their spatial claims on the sea, the convention aimed to regulate maritime conflicts and maintain post-World War II peace. However, it also drew up regulations for consolidating the borders of national Exclusive Economic Zones (EEZs) 200 nautical miles offshore and unleashed further border-making within them through the planning of these zones. The practice of intensified border manufacture and consolidation at sea as manifest by the EEZs is part of intensifying urbanisation processes with differing degrees of enclosure in practice (Fairbanks et al., 2018). These explicit forms of territorial regulation were conceived by Lefebvre as one of the three dimensions of the urban (Lefebvre 1991). The industrialisation of the maritime industries and the intensified use of sea space that advanced rapidly after the World War II marked the moment when the urbanisation of the sea began in earnest (Couling, 2023).

The Exclusive Economic Zones signified the formation of a radical new state space, described by Campling and Colas as 'the single greatest enclosure in human history' (Campling & Colás, 2017: 780). However, this enclosure—within which EU directives demanded that marine spatial plans should be delivered by coastal nations by 31 March 2021—is also a grey zone. While coastal nations command over the use of resources, they still do not have sovereign rights, meaning that the EEZs have become a mutated and compromised version of the sea commons (Couling, 2020a).

I have argued elsewhere that this type of urbanisation demanded more than accelerating industrial processes; it required a full state apparatus to transform the very nature of the sea space, render it abstract rather than 'natural', and then perform operations of homogenisation (Couling, 2023). This meant fundamentally transforming a complex, differentiated, living, and contingent space, resulting in one of the most contradictory examples of abstract state space in the Lefebvrian sense. For Lefebvre, state space displays 'a rationality of unification ... used to justify violence', a unification that tends towards homogeneousness, which is subsequently perceived as a consensus by the public (Lefebvre, 1991: 282). The global dimension of this new type of sea space became an historic precedent, arguably exemplifying the first rules of modern international law (Theutenberg, 1984). It also illustrates the advancing spatial control of the state, ostensibly aligned to the practical purpose of keeping order in the world's oceans and seas.

But events in contemporary seas demonstrate the opposite, resonating with Lefebvre's theory. The 'liquid violence' of oceanic borders—borders which are elemental in their composition and whose fluidity is harnessed politically—is most palpable in their enforcement on migrants in the Mediterranean by EU member states, or as reported by Forensic Oceanography, by the EU supporting the Libyan Coastguard to carry out the illegal task of 'refoulement' on their behalf (Heller & Pezzani, 2018, see also Şarlak, this volume). Ongoing Russian military exercises in the Barents Sea to demonstrate military power are a further example of the instrumentalisation of abstract borders to sharpen the profile of territorial political ambitions (Nilsen, 2023).

This chapter examines ways of thinking through borders and ocean governance in Norskehavet (the Norwegian Sea) to show that, while line-drawing is part of defining ocean boundaries, water masses are also spatially defined by temperature and salinity gradients across *oscillating* margins. Either narrow or broad, these margins are relatively constant, and can be measured and drawn according to oceanographic parameters. The chapter hence contrasts the layering of anthropogenic bordering practices across the sea's currents, depths, and water masses, with oceanographic principles. Based on a more-than-human perspective, explorations carried out by architecture students propose a loose framework of four zones that closely follow the Norwegian Sea's inherent physical, spatial organisation. Applying a morphology of oscillating margins would mean to align human activities more closely to the ocean's spatial and temporal rhythms. Within these categories (the spatial and temporal), two speculative projects exemplify more radical forms of counter-mapping that tackle the critical issues of salmon aquaculture and deep-sea mining, thereby submerging the observer and exposing the limitations to inherited conceptualisations of ocean borders writ large.

The Space of Borders

Ocean space is volumetric and kinetic—properties that linear, two-dimensional borders (as are often drawn on the map) are ill-equipped to reflect. While 'border' is loosely defined as a side or an *edge*, it can be geographical, political, or both (Oxford English Dictionary, n.d.). It also means *boundary*—the signifier of limits to territory among other things. In the case of the ocean, the concept of the border as an *edge* opens up a topographical, biological, and hydrographical space; the edge is not merely an abstract line, rather it has width, depth, landscape features, history, and protagonists. At the edge where sea 'meets' land, or liquid sea meets solid sea ice, there are transitional properties. For example, it is the sloping edge of the continental shelf that defines its geological limit at the oceanic crust, and therefore the extent of Exclusive Economic Zones. This volumetric edge condition is the site of water and nutrient exchange and is a space rich in biodiversity, providing vertical habitats for species such as cold-water corals. In northern seas and oceans, the ice edge—defined as where sea ice concentration is at least 15%—marks the beginning of the marginal ice zone which expands and retreats according to the seasons and with ice concentrations of up to 80%. The marginal ice zone in the Barents Sea can be as wide as 50 kilometres, and the location of the ice edge itself can vary between 50 and 250 kilometres monthly (Ingvaldsen, 2020). In spring, the ice edge releases valuable nutrients to the marine food chain. Sound artist Jana Winderen has described moving across this vast patchwork of spring ice melt in the Barents Sea, collecting walrus calls far beneath. Her sound compositions then bring these calls together with the distinct sounds of the ice itself, and the special forms of life within it (Winderen, 2018). Hence, oceanic edge spaces are highly productive, but mobile, fluid, and changeable, so tricky to determine.

And boundaries also have specific properties. In pre-modern times, when territory was less clearly defined, and control diminished with distance from centres of power, a boundary frequently signalled the beginning of a frontier region between two countries or districts called a 'march' or 'mark', which was normally overseen and defended by special arrangements (Chisolm, 1911: 689). This boundary had its own dimensions and characteristics, with gaps in between—recalling Faludi's criticism of the concept of territory as neatly parcelled national portions arranged 'shoulder to shoulder' (Faludi, 2018: 132). Instead, Faludi proposes thinking about states as islands in a sea of relations or even better, ice floats since they too change shape, grow, or diminish.

In his work on border theory, Nail defines the border as existing precisely 'between' states and emphasises the importance of understanding boundary phenomena and bordering as a process, rather than a static result. The border is also in motion—'made and remade according to a host of shifting variables' (Nail, 2016: 6). On land, geographical features such as rivers and mountain ranges have frequently been chosen to represent state borders—formations that are subject to uncontrollable change due to natural factors. The artist and research group Studio Folder have extensively explored and documented this phenomena steered by global warming and shrinking glaciers in the remote Alpine regions of Italy, Austria, Switzerland, and France in their project 'Italian Limes' (Pasqual et al., 2019), where national borders are ever sliding away from theoretical coordinates. At sea, Steinberg et al. investigated sea ice through a similar lens, and what the impact of an 'ice-sensitive' legal system in polar regions could be (Steinberg et al., 2020).

Since the 1494 Papal Treaty of Tordesillas, which divided the world between the two regimes of Spain and Portugal on either side of a line through land and sea, Western human interaction with the ocean in the world has been characterised by the attempted construction of borders, resulting in the drawing of primitive and approximate 'lines in the sea' (Peters, 2021) without geographical depth, without social 'concern',

(Latour, 2008) and without a full understanding of the complex, kinetic spatial system and habitat across which such a border was drawn. As a result, the 'superficial'—literally of the surface—oceanic border itself can only be a rough approximation of genuine difference or bifurcation. The edges, boundaries, and borders discussed above all expand their dimensions into broad *oscillating* zones. Closer examination of ocean space reveals a superimposed layering of borders defined by different parameters and interests, comparable to the territory as 'palimpsest' as proposed by Swiss urbanist and architectural historian André Corboz (1983).

Layers of Borders

Geophysical Borders

The International Hydrographic Organization undertook spatial delineation of the world's oceans and seas in 1952, in the common interests of sailors and mariners (International Hydrographic Organization, 1953). These limits were roughly based on bathymetry and had no political significance, rather aimed to share knowledge across different geographic regions and to standardise map-making practices. They are described through geographic coordinates that are still referred to today. According to this definition, the Norwegian Sea cuts a vast rectangle of water from the southern tip of Spitzbergen to Iceland across Jan Mayen Island, then eastwards to a latitude of 61 degrees North on the west Norwegian coast. The south-western limit corresponds to the geographic feature of the Greenland-Scotland ridge, and the western limit to the Mohn Ridge—part of the Mid-Atlantic ridge with depths of almost 4000 metres (see Fig. 1).

Biological Borders

Closely aligned to the geophysical borders outlined above, biogeographical boundaries describe species distribution. Science divides the world's continental shelves into seven climatic zones, further influenced by

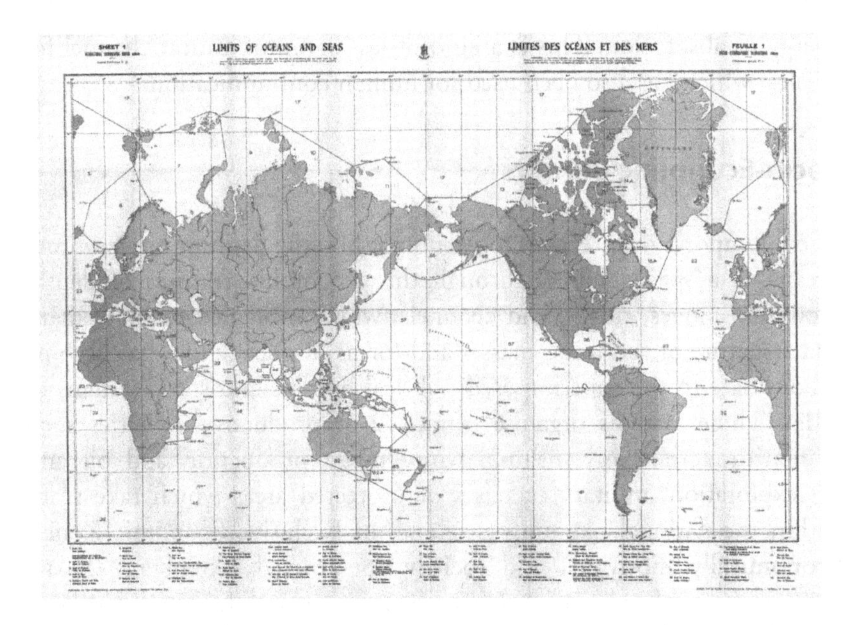

Fig. 1 Map accompanying second edition (1953) of IHO publication Limits of Oceans and Seas, Special Publication 23, with the Norwegian Sea marked number 6. Source: International Hydrographic Organization (1953: Sheet 1)

geographic conditions such as depth, resulting in 20 marine biogeographical regions.[1] These regions have no sharp borders, rather broad areas of transition between climate zones, in particular in the north–south direction. They are oscillating and 'dislocating' according to abiotic factors such as temperature and salinity levels (Brattegard & Holthe, 1995). According to this system, the Norwegian Sea is part of the East Atlantic Boreal Region. Marine species are adapted to habitats within these biogeographical zones, sometimes moving vast distances between them. Cutting through and across such biogeographical zones are spaces with specific characteristics, for example, concerning the transmission of sound—the SOFAR (Sound Fixing and Ranging) channel in the deep sea at depths of 800–1000 metres depending on temperature and salinity, where low frequency sounds can travel long distances without losing

[1] The seven climatic zones are arctic, boreal, northern temperate, tropical, southern temperate, anti-boreal, and Antarctic (Brattegard & Holthe, 1995).

energy to absorption. This is a kind of fast communication channel for whales, which has also been used for human communication.

Socio-Economic Borders

In contemporary western socio-economic life, the definition of seas and oceans as a space defined through the geographic features of depth, mountain ranges, ridges, and channels, which influence the movement and structure of the water masses and forms of marine life, has become subordinate to the superimposed political borders and static systems of order. These systems organise extractive or productive activities such as maritime transport, fishing, wind-energy production and oil and gas exploration, military exercises or protected areas which take place within the national planning envelopes of Exclusive Economic Zones. Economically motivated and socially constructed (Steinberg, 2001), socio-economic borders frequently extend far beyond the geophysical sea, as in the example of the Barents Sea (Couling, 2020b). In the case of the Norwegian Sea, the extent of the EEZ is not aligned to the geophysical conditions, rather stretches into the Greenland Sea west of Jan Mayen, since this island belongs to Norway. The 200 nautical mile limit east of Jan Mayen meets a 'banana hole' area of international waters—a 'left-over' slice of water between Norwegian EEZs to the east and west (Office of Ocean & Polar Affairs, 2020: 27) (See Fig. 2). However Norway's application to extend their EEZ and elimate this hole, was granted in recommendations by the Commission on the Limits of the Continental Shelf (CLCS). For planning purposes, a management area has then been defined by the Norwegian Environmental Agency, including the previous slice of international waters, and also not corresponding to the geophysical sea or the EEZ. These layers of borders both extend and fragment sea space. They define different seas according to specific agendas and regimes, while fragmenting ecological regions.

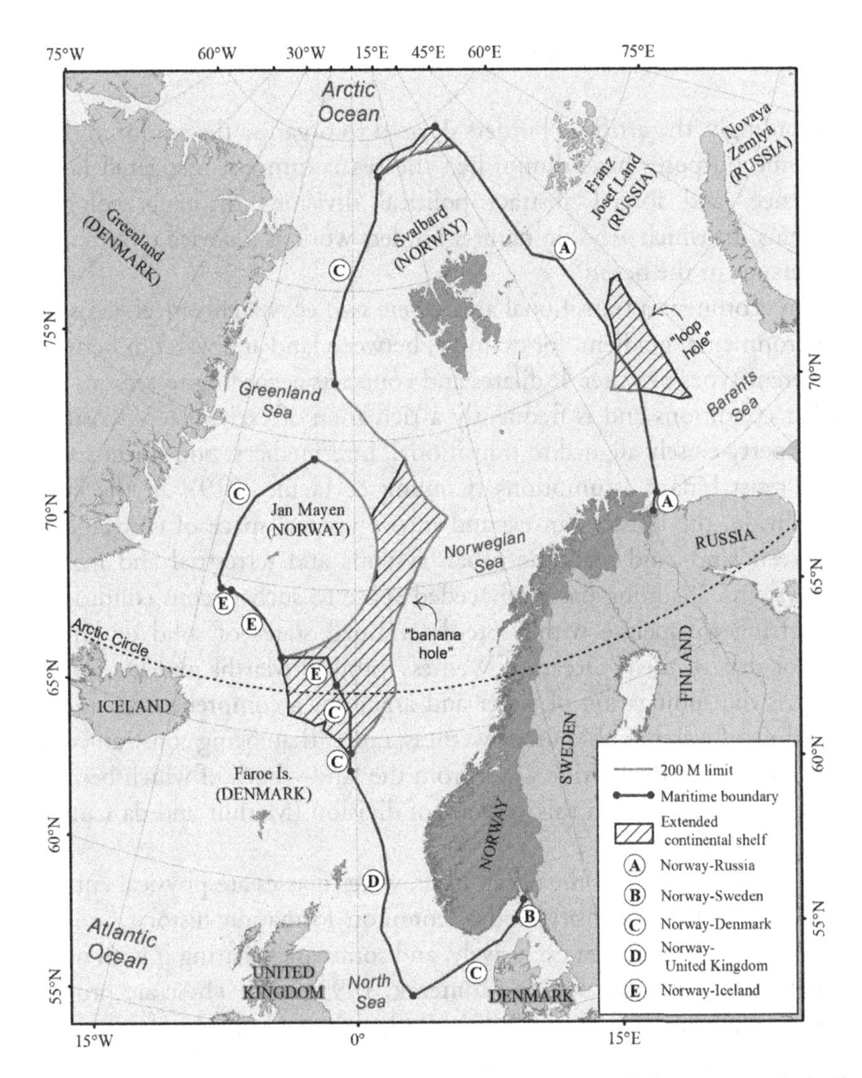

Fig. 2 Norway's claimed 200 nautical mile limits, extended continental shelf areas, and maritime boundaries, as submitted for analysis by the Office of Ocean and Polar Affairs, U.S Department of State (2020). Scale: 1:18,500,000. Source: United States Department of State Bureau of Oceans and International Environmental and Scientific Affairs (2020: 27)

Inherent Ocean Borders

Accordingly, the artificial borders devised to organise the sea for anthropogenic purposes have diminished the sea commons, instigated liquid violence, and forged abstract political divisions through ecological systems. In which ways do natural borders work as a device to articulate difference in the ocean?

An ecotone is a transitional area where two ecosystems meet across an environmental gradient, for example, between land and water or between different types of water. It dilates and contracts according to seasons and other conditions and is frequently a rich zone of exchange. Viscosity is a property closely aligned to transitional, intermediary, non-linear spaces that resist binary assumptions (Couling & Hein, 2019). At the land-sea divide, the border can expand into a viscous space of convergence between land- and seaborne urban systems and terrestrial and marine ecosystems. Planning has rarely ceded space to such viscous conditions, preferring to operate within pre-determined states of solid or liquid, wet or dry. In their 'Ocean of Wetness' project, Martha and Da Cunha discuss the subjugation of water and argue for a comprehensive understanding of water as ubiquitous wetness, rather than being constrained by 'designed' lines that divide water from the land—both of which become artificial states through this very act of division (Mathur and da Cunha, n.d.).

Within the space of the ocean itself, water masses are physical entities that have volume, density, and a 'common formation history',[2] sometimes occupying an area exclusively, and sometimes sharing it with other water masses through mixing (Tomczak, 1999: 279). These are properties normally associated with solids. Rather than a linear border, oceanic water masses are spatially defined by temperature and salinity gradients across oscillating margins. Either narrow or broad, these margins are also spaces with a specific (temperature/salinity) profile. Although fluid and rippling, these borders are relatively constant and can be measured and drawn according to oceanographic parameters.

[2] This means having its origin in a physical region of the ocean (Tomczak, 1999).

Flat, surface representations of the ocean therefore deny what cultural anthropologist Frank Billé calls its 'volumetric sovereignty' (Billé, 2020: 2). Examining such borders through counter-mapping could take us on a different journey through powerful edges, thresholds, zones, and volumes, which can only be discerned through an interior understanding of oceanic systems and are little explored in spatial terms. Architectural interpretations of the spaces usually described by ocean science are a form of counter-mapping of inherent ocean borders. It can offer an insight into a pulsating system that will potentially determine our life on a finite planet close to tipping point. Land- and surface-based territorial models used in marine spatial planning that rely on linear bordering practices have so far not been able to arrest the continued deterioration of the ocean's ecological condition. As we face limits of wholly new dimensions, which pose a seemingly unsurmountable challenge for ocean governance and planning, new perspectives will be required.

Counter-Mapping in the Norwegian Sea

Norwegian Sea

The Norwegian Sea is deep—on average reaching 2000 metres and at its deepest, just under 4000 metres. The English translation of its Norwegian name *Norskehavet* is literally the Norwegian *ocean*, a name that captures this extreme vertical dimension as well as its vast horizontality covering an area of 1,383,000 square kilometres. Oceans are distinguished from seas by the continental slope that reaches to the deep ocean floor. In oceanographic terms, a 'sea' would normally be part of the continental shelf, comprising 'the submerged prolongation of the land mass of the coastal State', under relatively shallow water (UN, 1982: 53). This spatial definition has become increasingly scrutinised since UNCLOS and the race for access to resources within exclusive economic zones.

A three-dimensional view of the ocean floor therefore reveals continental shelves spreading far beyond the familiar protruding landmasses in world maps. Circulating past these landmasses, the ocean's water masses are driven by currents, influenced by bathymetry, temperature, salinity, and wind, creating unique volumes of water that offer habitats to different forms of life. Three main water masses comprise the Norwegian Sea: cold Arctic water moving southwards from the Arctic Ocean, warm, salty Atlantic water as an extension of the Gulf Stream moving northwards with secondary currents turning inwards around basins and ridges, and less saline coastal waters from the Baltic and freshwater runoff from the mainland moving northwards close to the Norwegian coastline (Possenti, 2021) (see Fig. 3). Water masses are also vertically overlaid; the Norwegian Atlantic current (Atlantic water) occupies the top 100–600 metres in the Norwegian basin, and the cold deep-sea water occupies the water column from 1000 metres depth to the bottom. The intermediate layer of fluctuating thickness is called Arctic Intermediate water (see Fig. 4). In addition, the Greenland/Norwegian Sea is one of the five locations where thermohaline circulation causes deep water formation through the cooling and sinking of surface water (Rahmstorf, 2006). The Great Ocean Conveyor Belt circles through the world ocean, then overturns its water masses from top to bottom at these dedicated locations.[3] It drives global climate and is currently being observed slowing down due to climate change with unpredictable results on the global climate itself (Li et al., 2023).

Based on the Norwegian Sea's 'inherent' borders, the following section discusses a range of explorative methodologies open to engaging with the ocean's own spatial logic, as carried out by master students at Bergen School of Architecture (BAS).[4] By engaging with volumes, fluidity, relationality, and inter-species dialogue, the projects aim to enrich our spatial understanding and thereby also link the field of architecture to crucial questions embodied in the liquid volumes just one step over the BAS quay and into the fjord—across the artificially constructed threshold between land and sea.

[3] The five locations are the Greenland-Norwegian Sea, the Labrador Sea, the Mediterranean Sea, the Weddell Sea, and the Ross Sea (Rahmsdorf, 2006).

[4] Design studio 'Explorations in Ocean Space', 2021.

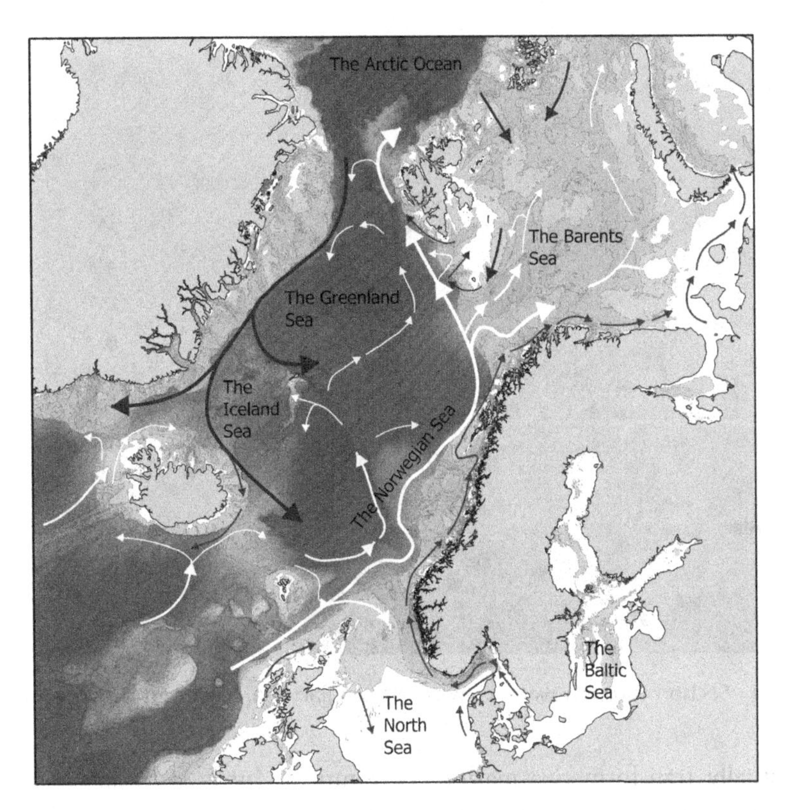

Fig. 3 Map of the major ocean regions treated in the report 'Observed and expected future impacts of climate change on marine environment and ecosystems in the Nordic region'. Red arrows show warm currents, blue arrows show cold currents. Coastal currents are shown in green. Source: Ottersen et al. (2023)

Method

Through a quasi-scientific method of registering oceanic life-forms and habitats, the investigation of Norskehavet's oscillating margins aimed to decipher this ocean's own internal system, spaces, and borders, and how a typological vocabulary of ocean space could be expanded, translated from science, and aligned to architectural thinking. In this way, conditions for counter-mapping, the *unmaking* of borders, and the tracing of oscillating margins could potentially be set up. Oceanic life and human-made artefacts have long since shared a mutually constituted and

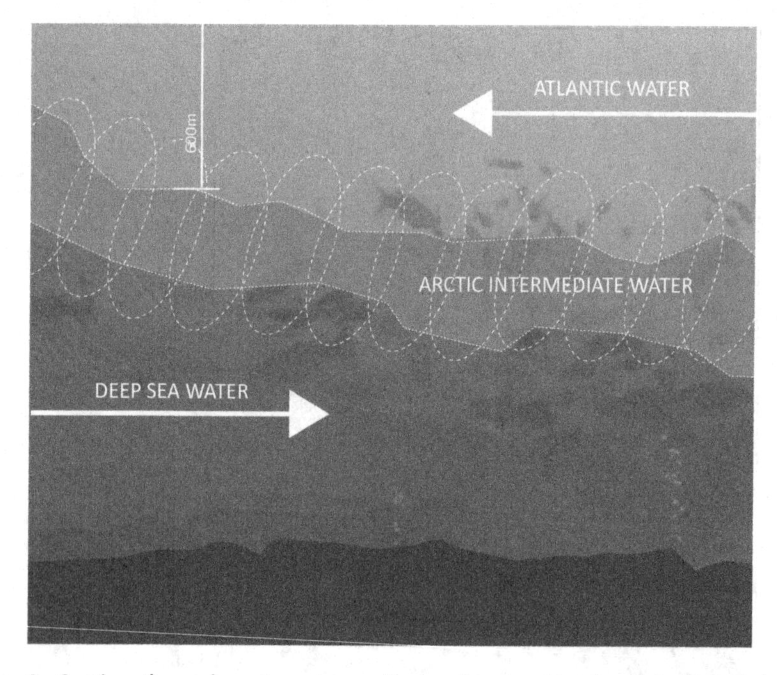

Fig. 4 Section through water masses, Norwegian Sea. Produced by the author

mutually transformative marine environment. Inspired by more-than-human perspectives and the absorption of the obsolete 'monsters' of the past industrial age into design considerations (Latour, 2011), the registering process included biotic protagonists, abiotic protagonists (waves, currents) that exert a strong influence on species habitats, and abiotic human-introduced invasive 'species' and activities such as oil rigs and trawling.

For spatial research on the ocean, territory as described by land-based divisions in adjacent zones is inadequate (Peters, 2020). Scales are both embedded and exponential, characteristics are overlapping, migrating, and transforming. A wholly new (for the Western world) conceptualisation of space would be required to capture these dimensions and dynamics—a contradiction that, for example, in indigenous oceanic understanding of the Polynesian or Sámi cultures, does not exist from the outset. Sámi spatial practices and understandings are characterised by fluidity between architecture and landscape, land- and seascape, physical and metaphysical elements, and different temporal/environmental conditions—they embrace situated and radically relational ways of knowing (Joks et al., 2020: 305). By both shifting the lens to speculate about 'borders' as experienced by different species and focusing on ecosystem interactions, three broad, alternate, Norskehavet zones emerged: Everywhere, the Continental Shelf, and the Deep Ocean as described below. In reference to the different regimes of bordering previously outlined in this chapter, these zones are proposed as an interpretation of 'inherent ocean borders', each containing innumerable spaces and conditions within them, yet also exhibiting a series of defining characteristics.

The Everywhere Zone

The spatial category 'Everywhere' is a characteristic of sea space that defies anthropogenic bordering since certain forms of marine life cannot be restricted to any one zone. Everywhere includes phyto- and zooplankton that form the basis of the marine food chain, currents, changes in temperature and pollution (see Fig. 5). Microplastics and plastic litter can literally be found everywhere in the ocean and defy localised efforts of control. The borderless ocean asserts the urgency of understanding the borderless effects of human actions, regardless of abstract economic or management zones. Everywhere redistributes cause and effect over vast distances into remote contexts or equally brings effects of distant actions close to home. Due to the system of Norwegian Sea currents, bays along the coastline of the Lofoten archipelago—a popular tourist destination renowned for its stunning land—and seascape and traditional fishing industry—are subject to high levels of plastic litter deposition (Solbakken et al., 2022).

The Everywhere zone intersects with specific local conditions, weaving through the other areas outlined below. Despite contemporary globalised lifestyles and consciousness, bordering practices have increased (Nail, 2016) and the resulting divisions have become deeply entrenched in spatial thinking. Industrialised society has difficulty grasping borderlessness as a spatial category of effect. The Norwegian Sea is an 'open' sea that exchanges water with its neighbouring seas and contains several internal

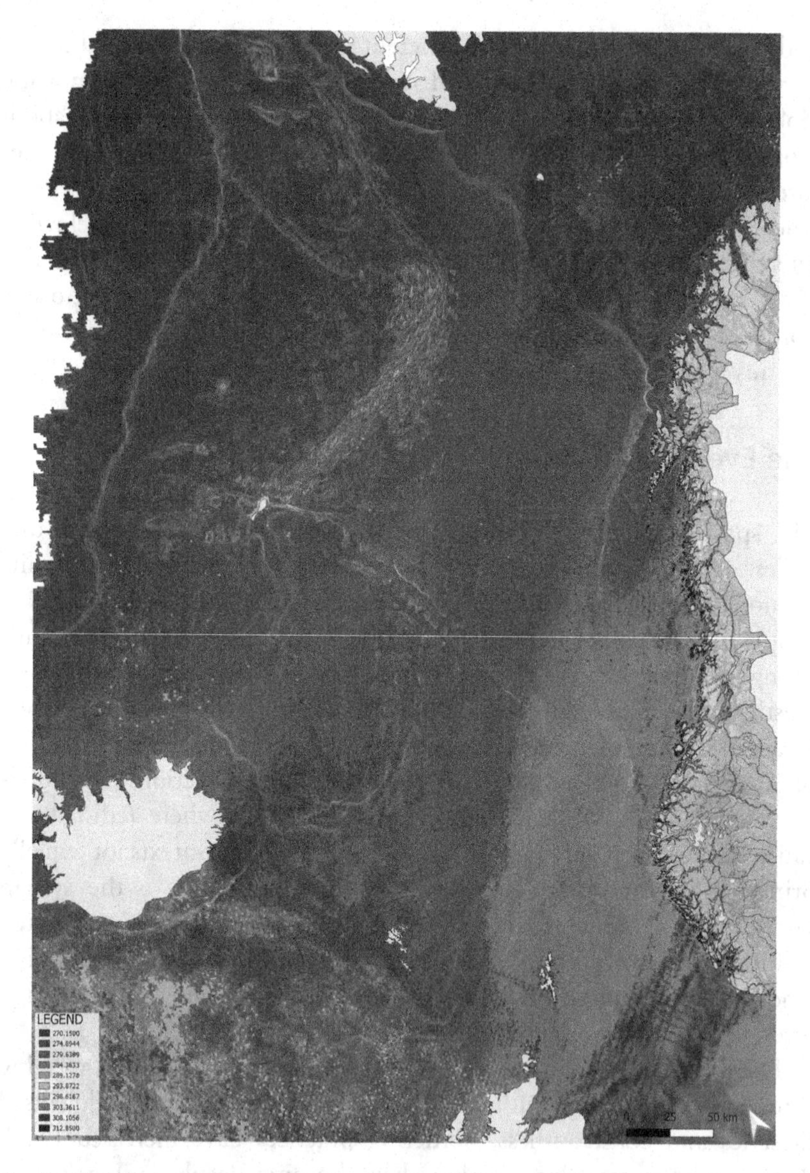

Fig. 5 The Everywhere zone: Temperature change in the Norwegian Sea. Source: Sashant Tiwari, BAS 2021

basins—it is not surrounded by land as in the case of the North or Baltic Seas. Therefore, the Everywhere zone also extends across the geophysical borders of the Norwegian Sea into the world ocean.

The Continental Shelf

Aquaculture, shipping lanes, oil fields, military installations, birds, spawning fish, trawlers, kelp, cold-water corals, and multiple marine species all inhabit the Norwegian Sea's continental shelf. Propelled by the warm Gulf Stream, it is a highly productive area ecologically, but is also the zone of the sea's most intense anthropogenic activity. It is fully urbanised. These activities are also mostly contained within the geophysical limit of the shallower waters and the continental slope; therefore, the proposed continental shelf zone is not new, rather gathers and reinforces existing geophysical and biological spaces and is based on ecosystem interactions (see Fig. 6).

The counter-mapping of the continental shelf articulates human interference throughout a zone defined by this area's oscillating limits. It is highly differentiated; therefore, four sites were selected that examine tensions and interactions in greater detail from a specific perspective: Runde (seabirds), Røst, (cold-water corals) Andoya (lost and resurgent mythologies), and Trondheim (salmon). The Trondheim site is used below as an illustrative example.

The Last Salmon

The Trondheim region is the original home of the Norwegian aquaculture industry. Since the first experiments on salmon smolt[5] by two brothers on the local island of Hitra in 1971, the industry has expanded exponentially to the multi-million kroner export earner it is today. At 50%, Norway is the world's largest producer of farmed salmon. Here, the aquaculture hub not only includes the world's largest sites for salmon farming, on Hitra and neighbouring Frøya, producing 260,000 tonnes

[5] Baby salmon.

Fig. 6 Activities on the Continental Shelf. Source: Sofya Markova, BAS 2021

of salmon per year, but also supports the industries around aquaculture construction, installation, supply, and tourism (see Fig. 7). Today, the highly technicised industry has little to do with Norwegian fishing traditions and is dominated by a handful of huge corporations—direct experience of the ocean itself is hardly required.

In Norwegian fjords, salmon-farming infrastructure is most frequently located towards the opening to the sea, but protected areas for salmon spawning are located deeper into the fjord. During its life-cycle, the path of an Atlantic salmon (*Salmo salar*) traverses the continental shelf both in the directions of the open ocean and the rivers feeding coastal fjords. This pathway is also an oscillating margin—moving up the river and back, defined by the sensory indicators of the smell of the river through

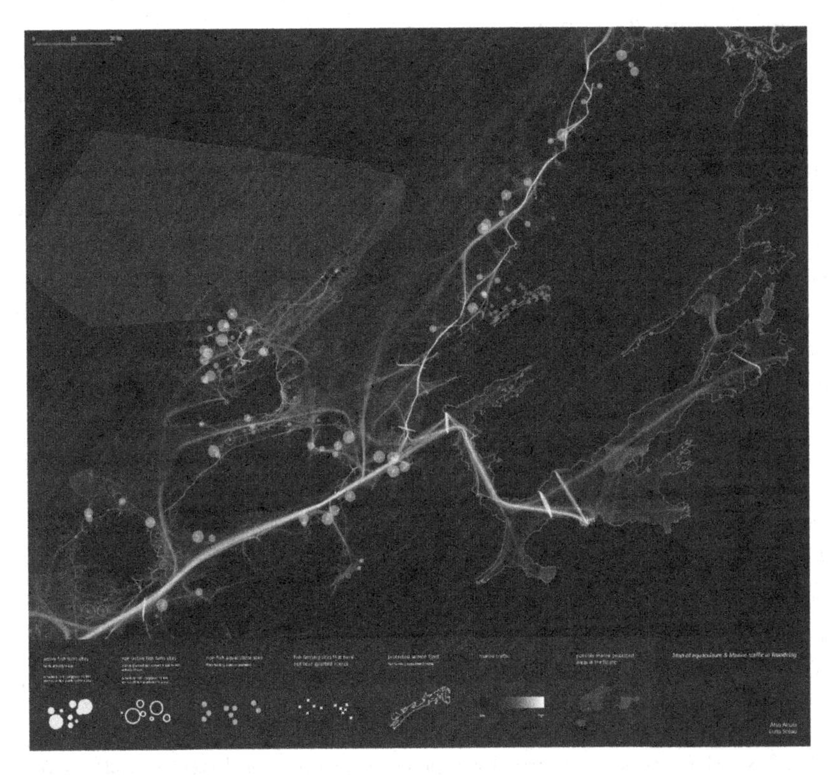

Fig. 7 Map of aquaculture and marine traffic in Trondelag. Source: Atso Airola and Luna Scéau, BAS 2021

which the salmon recognises the spaces of its home rivers and streams (Johnstone et al., 2012). Wild salmon moving towards spawning grounds further upstream must first successfully negotiate fish farms, endangered by the prolific levels of sea-lice prevalent there, the pollution caused by feed and excretion, and by the large numbers of escaped farmed salmon (Allen et al., 2019).

From the perspective of a salmon, how can invisible borders of lice infection, pollution, and escaped farmed salmon cutting across wild salmon paths to spawning grounds be defined? These are historic lines of connections comparable to trade or family connections in human societies. The cultural life of indigenous fishing communities in Norway, the coastal and river Sámi, is closely connected to the fjords, rivers, and their well-being; 'if salmon ceases to exist we would no longer be humans either' (Quinn, 2019: n.p.).

The counter-mapping project 'The Last Salmon' used the medium of a space-specific audio/video walk through an exhibition, inspired by the work of artist Janet Cardiff, to critically speculate on the acoustic conditions of salmon and the current state of aquaculture in Norway.[6] Sound travels five times further in water than on land. Hearing is one of the most important senses for fish and used for their navigation, therefore with increasing activity in sea space, sound levels in the ocean increase exponentially alongside stress levels for fish and other marine life. With the help of mobile phones and headphones, participants were visually and aurally guided through the installation following a series of pre-recorded events that enabled them to enter and navigate the experience of a wild salmon migrating. They became part of the salmon's journey starting from the river, swimming through the fjord, entering the open sea, and returning to the river. During this journey, the light changes. The salmon notices the turbid waters and finds it difficult to navigate. A boat can be perceived moving across the ceiling of the exhibition space—a double reality is created, where the fictional world of the film blends with the reality of the architecture and body in motion (see Fig. 8).

[6] The Last Salmon project was created and presented in Bergen in December 2021 by Atso Airolo and Luna Scéau, as well as the descriptive text assimilated in this paragraph. It received a first prize together with other project authors from this course in the universities section of the 2022 Lisbon Architecture Triennale in the category 'Multiplicity'.

Fig. 8 Double reality perceived through the life cycle of a salmon, superimposed into the exhibition space—the old Bergen jail—using a pre-recorded video accessed through mobile phones. Source: Atso Airola and Luna Scéau, BAS 2021

The Deep Ocean

An examination of the deep ocean plunges human perception into a zone of deep time and space—the zone below 200 metres where light is rapidly fading, where temperatures drop to an average of 4 °C, and where pressure increases to around 40–110 times that of the surface. The deep sea is defined by *vertical* borders that establish descending layers through the epipelagic, mesopelagic, bathypelagic, abyssopelagic, and hadopelagic zones. Here, the ocean floor has been formed through processes taking place over millions of years, resulting in spectacular geomorphic phenomena and forms of life that are still little understood.

Curving through Norskehavet, the Mid-Ocean ridge—a relatively recently discovered chain of volcanic mountains 65,000 kilometres long and up to 4200 metres deep encircling the globe—is where scientists estimate most of the earth's solid surface was born. Until the 1970s, it

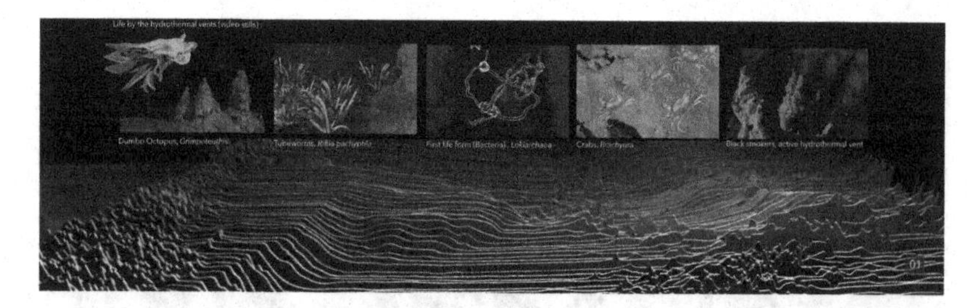

Fig. 9 Unique forms of life in the deep Norwegian Sea. Source: Katarina Kierulf, Bastian Haukefær, Sara Boukili and Maria Thoner, BAS 2022

was believed that life on earth depends on photosynthesis. But in the deep ocean, forms of life were discovered that develop through chemical processes—in particular, chemicals associated with hydrothermal vents expelling hot water up to a temperature of 340 °C from deep in the earth's crust (Ballard, 2000) (see Fig. 9).

The deep ocean provides 97% of the space available to life, yet humans cannot penetrate this space without immense technological and physical effort. Writer Jonathan Ledgard comments that this is not the space for humans, rather 'artificial intelligence sentinels, whose patience and fidelity will outstrip what we are capable of' (Ledgard, 2020: 62). The deep ocean layer has so far acted as its own protective border from human intervention.

Despite our scant knowledge of the unique deep ocean ecosystem, the International Seabed Authority who are responsible for regulating deep-sea mining in 'the Area'—outside of national jurisdiction and therefore part of the common heritage of mankind—are currently finalising regulations so that mining can start in the Clarion-Clipperton Zone between Hawaii and Mexico (Legal & Technical Commission, 2022). In Norway, technology developed for the petroleum industry is being directed towards deep-sea mining in Norskehavet using the narrative of renewable energy to justify this form of extraction, since minerals formed in the deep sea are sought after for use in batteries, wind turbines, and solar panels.[7] Governmental agencies have been surveying and sketching

[7] The list of minerals formed in the deep Norwegian Sea includes copper, aluminium, barium, silver, silicon, zinc, cobalt, manganese, gold, lithium, graphite, and nickel.

out a vast potential area for deep-sea mining, overlapping with valuable and vulnerable ecological areas (Alberts, 2023). Then, in January 2024, the Norwegian government voted to open up this area to mining, enabling companies to apply for licences to mine minerals including lithium, scandium, and cobalt (Bryant, 2024). Since these mining areas are inside the Norwegian EEZ, licenses can theoretically be issued in a short space of time. The following counter-mapping project, 'Deep Sea Dinner', was developed in response to this alarming prospect of deep-sea mining in Norskehavet, and witnessing of unbridled industry enthusiasm at a Deep-Sea Mining Conference.

Deep Sea Dinner

Welcome to an abandoned dinner party at 4000 metres below sea level,[8] where the atmosphere is dystopic, disorientating, conveying a feeling of deep uncertainty. The visitor is transported into an unknown world, where displaced consumable materials overlay future traces of human colonisation. Inspired by the irony of industry speeches at a deep-sea mining conference dinner, this installation brings the deep ocean close to human experience through spatial, audio, and visual means, exposing one of the last frontiers of planetary extraction at the same time.

Hydrothermal vents take minerals from magma, bringing them into the ocean as they smoke, where they then eventually fall as sediments to the sea floor. Meanwhile, the manganese crust is forming at a rate of few millimetres per million years. In the dining room installation, this magical, timeless world floats through the space—again heedless to human-conceived boundaries of speculative exploration licenses (see Fig. 9).

Deep-sea sediments and crust are being discussed as the new oil and are being explored with oil technology under the pretext of meeting

[8] The Deep Sea Dinner was presented in Bergen in December 2021 as an installation, sound-recording, and an accompanying magazine laying out background facts to the colonisation of the sea through borders and the deep-sea mining industry. This paragraph is an excerpt from their accompanying text. It was selected as a finalist in the universities section of the 2022 Lisbon Architecture Triennale in the category 'Visionaries'. Project authors: Katarina Kierulf, Bastian Haukefær, Sara Boukili, Maria Thoner.

green energy demands. This is the next chapter of Norway's financial odyssey; in July 2021, the Norwegian oil fund surpassed 12,000 billion kroners for the first time. The appropriation and consumption associated with deep-sea mining in the Norwegian Sea are interpreted in this project through an abandoned future dinner party where the sound track is still rolling—enthusiastic speeches of unimaginable profits overlaid by deep-sea sounds. A magnified view shows strangely coloured creatures, glowing minerals, and smoking vents. We are deep under the ocean—where we do not belong (see Fig. 10).

Fig. 10 Deep Sea Dinner project poster for the universities section of the 2022 Lisbon Architecture Triennale in the category 'Visionaries' (one of seven finalists), showing mining information and a view of the installation in the old Bergen jail, December 2021, which included sound and video projections. On the menu are Manganese steak, cables, gold, and tubeworms, washed down with a glass of vintage oil. Source: Katarina Kierulf, Bastian Haukefær, Sara Boukili and Maria Thoner, BAS 2021

Conclusion: Counter-Mapping Otherwise

The methods of counter-mapping discussed in the second section of this chapter are far-removed from conventional cartography, hardly utilising two-dimensional modes of representation, but rather exploiting sound, video, and three-dimensional installations to enable a sensual immersion into a submerged environment. The space of the sea is simulated to facilitate first-hand experience of the critical conditions found there and to provoke reflection. To tackle ocean governance beyond borders, inherited limitations in representing and experiencing ocean space must also be challenged. The work presented here does not assume to offer solutions to the complexity of marine spatial planning, rather it proposes a morphology of oscillating margins deriving from the ocean's own internal spatial logic, and therefore a deeper understanding of border conditions—both spatial and temporal—within which human activities could be more closely aligned. Everywhere, the Continental shelf, and the Deep Ocean can be understood as overlaying spatial categories in the Norwegian Sea, to which specific principles of utilisation, preservation, and protection could ultimately be applied.

References

Alberts, E. C. (2023, April 20). Norway proposes opening Germany-sized area of its continental shelf to deep-sea mining. *Mongabay Environmental News*. Retrieved June 15, 2024, from https://news.mongabay.com/2023/04/norway-proposes-opening-germany-sized-area-of-its-continental-shelf-to-deep-sea-mining/

Allen, S., Bankes, N., & Ravna, Ø. (2019). *The rights of indigenous peoples in marine areas*. Bloomsbury Publishing.

Ballard, R. (2000). *The eternal darkness: A personal history of deep-sea exploration*. Princeton University Press.

Billé, F. (2020). *Voluminous states: Sovereignty, materiality, and the territorial imagination*. Duke University Press.

Brattegard, T., & Holthe, T. (1995). *Distribution of marine, benthic macro-organisms in Norway: A tabulated catalogue (preliminary edition).* Norge Direktoratet for Naturforvaltning.

Bryant, M. (2024, January 9). Norway votes for deep-sea mining despite environmental concerns. *The Guardian.* Retrieved June 15, 2024, from https://www.theguardian.com/environment/2024/jan/09/norway-set-to-approve-deep-sea-mining-despite-environmental-concerns

Campling, L., & Colás, A. (2017). Capitalism and the sea: Sovereignty, territory and appropriation in the global ocean. *Environment and Planning D: Society and Space, 36*(4), 776–794.

Chisolm, H. (1911). March. In H. Chisholm (Ed.), Encyclopædia Britannica (11th ed., Vol. 17). Cambridge University Press. Retrieved June 15, 2024 from https://ia600208.us.archive.org/31/items/EncyclopaediaBritannicaDict.a.s.l.g.i.11thed.chisholm.1910-1911-1922.33vols/17.EncycBrit.11th.1911.v17.LOR-MEC..pdf

Corboz, A. (1983). *Le Territoire comme palimpseste.* Verdier.

Couling, N. (2020a). Extensions and viscosities in the North Sea. In N. Couling, & C. Hein (Eds.), *The urbanisation of the sea: From concepts and analysis to design* (pp. 189–203). nai010.

Couling, N. (2020b). Ocean space and urbanisation: The case of two seas. In N. Couling, & C. Hein (Eds.), *The urbanisation of the sea: From concepts and analysis to design* (pp. 189–203). nai010.

Couling, N. (2023). Losing sea. Abstraction and loss of the commons in the North Sea. In C. Schmid, & M. Topalović (Eds.), *Extended urbanisation. Tracing planetary struggles* (pp. 159–196). Birkhäuser.

Couling, N., & Hein, C. (2019). Viscosity. *Society & Space, Part 3: Turbulence.* Retrieved June 15, 2024 from http://societyandspace.org/2019/03/17/viscosity/

Fairbanks, L., Campbell, L. M., Boucquey, N., & St. Martin, K. (2018). Assembling enclosure: Reading marine spatial planning for alternatives. *Annals of the American Association of Geographers, 108*(1), 144–161.

Faludi, A. (2018). *The poverty of territorialism: A neo-medieval view of Europe and European planning.* Edward Elgar Publishing.

Grotius, H. (1609). Mare Liberum, sive de jure quod Batavis competit ad Indicana commercia dissertatio. *Lodewijk Elzevir.* Retrieved June 15, 2024, from https://upload.wikimedia.org/wikipedia/commons/7/7b/Grotius_Hugo_The_Freedom_of_the_Sea_(v1.0).pdf

Heller, C., & Pezzani, L. (2018). *Forensic oceanography. Mare Clausum. Italy and the EU's undeclared operation to stem migration across the*

Mediterranean. Goldsmiths, University of London. Retrieved June 15, 2024, from https://content.forensic-architecture.org/wp-content/uploads/2019/05/2018-05-07-FO-Mare-Clausum-full-EN.pdf

Ingvaldsen, R. (2020, February 29). The Barents Sea has become cooler, pushing the marginal ice zone south. *Institute of Marine Research News.* Retrieved June 15, 2024, from https://www.hi.no/en/hi/news/2020/january/the-barents-sea-has-become-cooler,-pushing-the-marginal-ice-zone-south

International Hydrographic Organization (IHO). (1953). *Limits of Oceans and Seas* (Special Publication Nr. 23). International Hydrographic Organization. Retrieved June 20, 2024, from https://epic.awi.de/id/eprint/29772/1/IHO 1953a.pdf

Johnstone, K. A., Lubieniecki, K. P., Koop, B. F., & Davidson, W. S. (2012). Identification of olfactory receptor genes in Atlantic salmon Salmo salar. *Journal of Fish Biology, 81*(2), 559–575.

Joks, S., Østmo, L., & Law, J. (2020). Verbing meahcci: Living Sámi lands. *The Sociological Review, 68*(2), 305–321.

Latour, B. (2008). *What is the style of matters of concern?* Uitgeverij Van Gorcum.

Latour, B. (2011). Love your monsters: Why we must care for our technologies as we doour children. In M. Shellenberger, & T. Nordhaus (Eds.), *Love your monsters: Postenvironmentalism and the anthropocene* (pp. 19–26). Breakthrough Institute.

Ledgard, J. (2020). Everything has its depth. In N. Couling, & C. Hein (Eds.), *The urbanisation of the sea: From concepts and analysis to design* (pp. 61–70). nai010.

Lefebvre, H. (1991). *The production of space* (D. Nicholson-Smith, Trans.). Blackwell.

Legal and Technical Commission. (2022). *Draft Regulations on exploitation of mineral resources in the area.* (ISBA/25/C/WP.1). International Seabed Authority. Retrieved June 15, 2024, from https://www.isa.org.jm/documents/isba-25-c-wp-1/

Li, Q., England, M. H., Hogg, A. M., Rintoul, S. R., & Morrison, A. K. (2023). Abyssal ocean overturning slowdown and warming driven by Antarctic meltwater. *Nature, 615*(7954), Article 7954. https://doi.org/10.1038/s41586-023-05762-w

Lien, V. S., Hjøllo, S. S., Skogen, M. D., Svendsen, E., Wehde, H., Bertino, L., ... & Garric, G. (2016). An assessment of the added value from data assimilation on modelled Nordic Seas hydrography and ocean transports. *Ocean Modelling, 99*, 43–59.

Mathur, A., & da Cunha, D. (n.d.). *Ocean of wetness*. Platform-learn-more. Retrieved November 14, 2023, from https://www.mathurdacunha.com/platform-learn-more

Nail, T. (2016). *Theory of the border*. Oxford University Press.

Nilsen, T. (2023, November 8). Northern Fleet kicks off major Barents Sea command and staff exercise. *The Independent Barents Observer*. Retrieved June 15, 2024, from https://thebarentsobserver.com/en/security/2023/08/northern-fleet-kicks-major-barents-sea-command-and-staff-exercise

Office of Ocean and Polar Affairs. (2020). *Limits in the seas. Norway- Maritime claims and boundaries* (148; Limits in the Seas). Bureau of Oceans and International Environmental and Scientific Affairs, U.S. Department of State. Retrieved June 15, 2024, from https://www.state.gov/wp-content/uploads/2020/08/LIS148-Norway.pdf

Ottersen, G., Børsheim, K. Y., Arneborg, L., Maur, M., Schourup-Kristensen, V., Rosell, E. A., & Hieronymus, M. (2023). *Observed and expected future impacts of climate change on marine environment and ecosystems in the Nordic region* (2023–10; Rapport Fra Havforskningen). Institute of Marine Research. Retrieved June 3, 2024, from https://www.hi.no/en/hi/nettrapporter/rapport-fra-havforskningen-en-2023-10

Pasqual, E., Bagnato, A., & Ferrari, M. (2019). *A moving border: Alpine cartographies of climate change*. Columbia Books on Architecture and the City.

Peters, K. (2020). The territories of governance: Unpacking the ontologies and geophilosophies of fixed to flexible ocean management, and beyond. *Philosophical Transactions of the Royal Society B: Biological Sciences, 375*, 20190458. https://doi.org/10.1098/rstb.2019.0458

Peters, K. (2021). A line in the ocean 30×30: Ocean borders and geography's limits. *Chair's Plenary Lecture at the Royal Geography Society (with IBG) Annual International Conference*, London, August 31–September 3, 2021.

Possenti, L. (2021). *Autonomous ocean carbon system observations from gliders* (Doctoral Dissertation). University of East Anglia.

Quinn, E. (2019, November 2). What a Saami-led project in Arctic Finland can teach us about Indigenous science. *Eye on the Arctic*. Retrieved June 15, 2024, from https://www.rcinet.ca/eye-on-the-arctic/2019/02/11/what-a-saami-led-project-in-arctic-finland-can-teach-us-about-indigenous-science/

Rahmstorf, S. (2006). Thermohaline ocean circulation. In S. A. Elias (Ed.), *Encyclopedia of quartenary sciences*. Elsevier.

Selden, J. (1663). *Mare Clausum: The right and dominion of the sea in two books*. Printed for Andrew Kembe and Edward Thomas.

Solbakken, V. S., Kleiven, S., & Haarr, M. L. (2022). Deposition rates and residence time of litter varies among beaches in the Lofoten archipelago, Norway. *Marine Pollution Bulletin, 177*, 113533. https://doi.org/10.1016/j.marpolbul.2022.113533
Steinberg, P. E. (2001). *The social construction of the ocean.* Cambridge University Press.
Steinberg, P., Kristoffersen, B., & Shake, K. (2020). Edges and flows: Exploring legal materialities and biophysical politics of sea ice. In I. Braverman & E. Johnson (Eds.), *Blue legalities: The life and laws of the sea* (pp. 85–106). Duke University Press.
Theutenberg, B. J. (1984). Mare Clausum Et Mare Liberum. *Arctic, 37*(4), 481–492.
Tomczak, M. (1999). Some historical, theoretical and applied aspects of quantitative water mass analysis. *Journal of Marine Research, 57*(2), 275–303.
United Nations (UN). (1982). *UNCLOS 1982.* Retrieved June 15, 2024, from https://www.un.org/depts/los/convention_agreements/texts/unclos/unclos_e.pdf
Winderen, J. (2018, November). *Spring bloom in the marinal ice zone* [CD, digital]. Touch – Tone 65.

Bordered-In, Bordered-Out, and Overlapping Territorialities in Ocean Space: The Case of Fisheries

Po-Yi Hung

Modern nation-states have strategic interests in maritime borders for resource governance, territorial control and national security (Hung & Lien, 2022). Since 1958, the United Nations has promulgated a series of agreements on the oceans (Alexander, 1968). The 1982 United Nations Convention on the Law of the Sea (UNCLOS) and its amendments, including the 2023 binding treaty on Marine Biodiversity in Areas beyond National Jurisdiction (BBNJ), now frame most of these state intentions. UNCLOS divides the ocean into specific spaces with legal rules to govern it. This represents spatially-based (inter)national political interventions in modern ocean use (Braverman & Johnson, 2020; Lambach, 2021). This global normative framework extends state sovereignty to the ocean and governs marine resource extraction and environmental conservation.

Social science disciplines have long been concerned about the political geography of, and geopolitics related to, marine resources and the

P.-Y. Hung (✉)
National Taiwan University, Taipei, Taiwan
e-mail: poyihung@ntu.edu.tw

K. Peters and J. Turner (eds.), *Ocean Governance (Beyond) Borders*,
https://doi.org/10.1007/978-3-031-71322-4_4

environment (Couper, 1978; see also Benjaminsen et al., 2017). Indeed, 'resources' and 'environment' are the two major themes through which ocean borders and the maritime governance of the state have been conceptualised (Hung & Lien, 2022). Scholars have explored various bordering processes of ocean space aiming at regulatory efficiency and equity of use, not least for economic purposes (see, e.g., Østhagen, 2020; Zaucha & Jay, 2022). Even though environmental protection and resource extraction are discussed in relation to one another—for example, with protection meant to 'manage' resource extraction—they nevertheless appear to be at odds with one another, frequently because of competing interests. Yet both require and rely on boundary demarcation to achieve their respective purposes. Allocating space for economic use often relies on boundaries. So too does allocating space for conservation practices. Therefore, marine resource extraction and environmental protection are intricately related to ocean borders/borderings. They are both ways in which state governance over maritime space is realised.

Fisheries are a topic at the very intersection of the related spatial conflicts between resource extraction and environmental protection of the ocean. On the one hand, the development of the fishing industry worldwide relies on the increasing catch of the fish stock to make maximum profits. Managing 'catch' has partly depended on establishing fishing quotas, allocating fishing zones and other regulatory measures. On the other hand, the depletion of the fish resources meanwhile entails actions concerning resource conservation for the sustainability of the global fisheries. Indeed, in recent political ecology studies that focus on the relationship among ocean borders, resource frontiers and state power, fisheries are one of the topics that have received attention (Boucquey, 2020; Carothers, 2011; Carothers & Chambers, 2012; Fabinyi et al., 2015; Knott & Neis, 2017; Satizábal & Dressler, 2019). The local, national and global scales are distinct but connected, and they are the basis for discussions about the intricacies of the fishing business and the relationships between use and abuse (of what, by whom).

At the global and national scales, research on the fisheries and the resource governance regime focuses on spatial demarcation, such as territorial seas and Exclusive Economic Zones (EEZs), emphasising state sovereignty, territorial privatisation and the expansion of national sovereign rights (and private enterprise) over the marine resources (Campling & Colás, 2018; Campling & Havice, 2014; Fairbanks, 2019). Indeed, the creation of EEZs as part of UNCLOS saw the largest enclosure of maritime space and 'they have served as the foundations for a turn toward enclosure through rights- and market-based regimes in the sector' (Fairbanks et al., 2018: 148). At the local and community scales, many studies, alert to power asymmetries, justice and the violence of modes of extractive capitalism, focus on the interactions between the state and Indigenous peoples or small-scale, artisanal fishing people; especially the actions and attempts of these local actors to obtain non-state territorial claims over waters that are theirs (Bennett et al., 2015; Foley & Mather, 2019; Pinkerton & Davis, 2015; Satizábal & Dressler, 2019).

These studies, conducted at various scales, have led to the primary understanding of the ocean as a complex area where fish resources are linked to understanding human and environmental exploitation, as well as the protection of human rights (though these are frequently incompatible; see Figueroa et al., 2023; TNI Agrarian Justice Programme et al., 2014). The territorialisation process of maritime spaces is linked explicitly to the control of fish resources, the practice of state governance and legal rationality and authority. Here ocean *borders* and bordering practices are crucial. Not only are they at the core of control processes, but they are also themselves reconceptualised through the practices of fishing activities in different contexts (Hung & Lien, 2022).

This chapter draws from the example of fisheries to unpack the politics of ocean borders in relation to both resource extraction and environmental protection and in view of state territoriality. For analytic purposes, I divide the content into three sections: bordered-in territoriality, bordered-out territoriality and overlapped territoriality. The first section, *bordered-in territoriality*, focuses on fisheries and conservation within the territorial seas and EEZs, where ocean borders are associated with state's sovereign rights. Second, I take the space of high seas as the *bordered-out territoriality*, emphasising the high seas as the commons and the ways the state performs its territoriality without claiming sovereign rights via distant-water fisheries. Third, I use *overlapped territoriality* to address the disputed maritime grey zones between different states and how fisheries have played a key role in bordering-up a cooperative space of sovereign rights for fishing.

To note, although I propose three themes in three sections respectively, it is not my intention to claim that the three sections here will comprehensively cover all the dimensions regarding relations between fisheries, ocean borders and (state) territoriality. Rather, my purpose is to use fishing practices across space and scales as the starting point to disclose a variety of and to complicate the relationship between, ocean borders and state power in governing the maritime space. Meanwhile, I do not intend for the three sections to be independent from each other. On the contrary, the scenarios mentioned in the following three sections are in fact interconnected with each other. That said, ocean borders, as maritime lines, are not only dividing but also connecting the demarcated parts of the ocean. While ocean borders intend to *cut up* the ocean into different spaces, the ocean is physically de facto a huge, planetary waterbody. As Steinberg and Peters note (2015), the ocean is mobile, churning, voluminous and requires us to call into question ways of governing. As a result, bordered-in territoriality, bordered-out territoriality and overlapped territoriality over the ocean actually influence and interact with each other in view of the fishing activities through the very physical qualities of water itself.

Finally, although this chapter focuses on discussions about fisheries, the critical points addressed actually extend beyond this activity and connect to debates regarding other marine resources in general. In other words, in addition to fisheries, a variety of resource excavation or extraction, such as deep-sea mining (Childs, 2019; Carver et al., 2020; Peters et al., 2018) and the offshore wind farms (Hung, 2020), all also entail bordering actions and territorialisation over the ocean. As such, ideas concerning the bordered-in territoriality, bordered-out territoriality and overlapped territoriality apply likewise to marine resources other than fish and allow concepts for critically investigating state governance over the ocean in other scenarios, too, I next explore the three themes highlighted in this introduction in greater detail.

Bordered-In Territoriality: Inside State Sovereign Rights

By *bordered-in territoriality*, I refer to the bordering actions within the territorial seas and EEZs. These are ways of territorialising the oceanic space either for securing (more) fishing resources and/or for protecting the marine resources and oceanic ecosystems in general. One of the most active research frameworks in understanding the ocean and its bordered-in qualities is when the state engages it as a resource frontier (Fawcett et al., 2022). In this view, the ocean is similar to a land-based resource frontier, extended by the logic of territorialisation of state power (Rasmussen & Lund, 2018). I take fisheries as an epitome of the process. This means fisheries at first reconfigure the oceanic space as a resource frontier. As Havice and Zalik note,

> ... the oceans ... are often described as the last planetary frontier. This is a metaphor which like frontier thinking applied to terrestrial space, uncritically divides the human from "wilderness" and leaves "the frontier" as a boundary that remains intact and out of human reach. In western historiography, the frontier has been associated with terrestrial projects of violent conquest and colonialism, racism, imperialism, and resource-fuelled global capitalism, implying a boundary to be breached, controlled and civilised. Thus, frontiers combine the creation of commodities with cultural and territorial control, making a range of natural and social processes available for appropriation. (2018: 220)

Indeed, fisheries industries trigger actions of the state in territorialising the resource frontier of the ocean in terms of the establishment of rules and laws for governance, for resource extraction (but also for conservation). In other words, the state has never stopped its intention to territorialise the resource frontier of the ocean. Rather, as Havice and Zalik (2018) point out, the border of the oceanic frontier is one to be breached and then appropriated, brought into the 'fold' of territorial space.

Political ecologists have long examined how the state territorialises the resource frontier via fisheries and have engaged critical discussion around this process and *against* the blue economy on which this territorialisation rests. Under the discussion of the contemporary blue economy (Voyer et al., 2018; see also Mulazzani & Malorgio, 2017), marine resource extraction is characterised by the commodification with the value production of nature and the ensuing demarcation of ocean borders (Ertör & Hadjimichael, 2020). This approach also rests on another border—between 'culture' and 'nature'—with economic uses or 'services' resting upon an understanding of the environment, or ocean, as a separate sphere providing for life, rather than entangled with it. This false binary is highlighted particularly in Indigenous scholarship on the ocean (see George & Wiebe, 2020).

In critiquing the 'bordered-in' territory of oceanic control, geographers and other scholars in critical ocean studies have noted a shift towards a more planned economy, allocating ownership rights as well as establishing mechanisms and systems to safeguard the state interests and legal rationality (Voyer et al., 2018; see also Winder & Le Heron, 2017), or arguing for modes of blue 'degrowth' (Hadjimichael, 2022). However, other critical work also pushes these debates further. Political ecologists have explored the parallels between land- and ocean-grabbing through a long-standing focus on power relations between environmental issues and socioeconomic structures. Although the concept of blue economy conveys the idea that growth and development have been achieved through a rational division of the ocean, political ecologists and other social scientists challenge the state's territorialisation process of ocean resources through the critical perspective of *ocean grabbing* (Barbesgaard, 2018; Bennett et al., 2015; Foley & Mather, 2019).

The concept of ocean grabbing often means actions, policies or programmes of the state (or capital) that deprive small-scale fishery people of their resources, seize coastal lands from vulnerable communities and undermine historical ocean access rights (Bennett et al., 2015; TNI Agrarian Justice Programme, Masifundise and Afrika Kontakt, 2014). Ocean grabbing is a contemporary state's act of allocating space to achieve resource enclosure (Foley & Mather, 2019). In order to highlight the growth of the blue economy and blue growth as well as the desire to grab more, academics studying the politics of ocean grabbing thus consider the discourse and ideology surrounding these concepts

(Barbesgaard, 2018). Studies of ocean grabbing, therefore, show the ocean as a violent space of conflict and dispossession (though note that Foley and Mather also highlight that 'local users' of oceans are not passive or inert to state pressure but resistant too, and can 'grab' or 'grab back' resources [2019: 298] or as Figueroa et al. [2024] note, seek justice when prosecuted for engaging territories to which they have rights). In particular, political ecologists have politicised the analysis of ocean borders, not least in relation to territorial seas and the EEZ through attention to power and politics, narrative and knowledge, scale and history, environmental justice and equity (see Bennett [2019] for a review).

Meanwhile, ocean grabbing is not just about bordering the ocean for the removal of resources, such as fish. In addition, ocean grabbing is also a bordering action for environmental conservation to (apparently) sustain the fishing resources or to 'protect' specific marine ecology. Through the spatial logic of enclosure and privatisation, the governance of the ocean as a resource frontier has coincided with the state's intervention in establishing marine protected areas (MPAs) in the name of environmental protection (Fairbanks et al. 2018). Coupling with environmental protection and state territorialisation, natural conservation has long been viewed as a practice of state power (Neumann, 2004; Peluso, 1993; Tafon, 2018). For example, political ecologists have highlighted the use of state violence in conservation programmes, even in armed, quasi-war situations (Duffy, 2014; Dunlap & Fairhead, 2014; Peluso & Vandergeest, 2011; Vandergeest & Peluso, 1995). Violence thus becomes a central theme in explaining the reproduction of state power and territorial politics of natural conservation (Bocarejo & Ojeda, 2016; Peluso & Watts, 2001; Ybarra, 2016).

Ocean bordering has been part of these critical discussions (Kamat, 2018; Raycraft, 2019; Stern, 2008). Regarding conservation violence and related national territorialisation, a political ecology approach provides an analytical framework for exploring ocean bordering and marine conservation. For example, Muralidharan and Rai (2020) used the Gulf of Mannar Marine National Park in India as an example to highlight the marine context of poaching and trafficking activities as well as their conflict with marine conservation. From Muralidharan and Rai's (2020) perspective, these conflicts disclose how the state seized, or grabbed, the oceanic space from local artisanal fishermen communities through the alleged claims for marine conservation and the associated bordering actions. Elizabeth De Santo's (2020, 2024) works have demonstrated how states also establish 'conservation territories' (see also Gray, 2018) to grab space under the guise of conservation, but to in fact establish geopolitical footholds in strategic oceanic spaces.

To sum up for this section, political ecology and political ecologists in particular (alongside critical geography and marine social science), as addressed above, have provided avenues into critical investigations as well as reconsiderations regarding the relations among fisheries, ocean borders and state territoriality. More specifically, I have used the term *bordered-in territoriality* to highlight the process through which fisheries have been a way for the state to turn the ocean into a resource frontier and then further transform it into a bordered territory of the state in the name of resource extraction or protection—both obtained through processes of 'grabbing'. To note, territorial seas and EEZs embody different state power. Territorial seas are deeply associated with state sovereignty. In EEZs the state can only claim its sovereign rights over marine resources *without* practices of sovereignty. They cannot claim the space as legal territory, but they can nonetheless territorialise through practices and processes of spatial allocation. Specifically, both territorial seas and EEZs are spaces where the ocean border also conveys the state power with regulations and laws. In other words, borders become the mechanism for realising the state territorialisation for oceanic governance purposes. However, the discussions above also raise another question concerning the situation of the high seas. While the high seas are recognised as the international commons without the bordering actions from the state, how does the state perform its territoriality via fisheries? This is the question this chapter will next proceed with.

Bordered-Out Territoriality: Beyond State Sovereign Rights

While echoing political ecologists' approaches to ocean grabbing, this section also proceeds with caution, not least in view of discussions of marine resource acquisition that note its territorial politics as largely aligned with *terrestrial* models of capital accumulation in *terrestrial* political economies (Foley & Mather, 2019). Indeed, many argue that the modes of governance at sea lend from the land (Lambach, 2021; Peters, 2020a, 2020b). Yet land-based ocean governance frameworks raise another concern about the transformation and implementation of *high seas* governance, since the high seas do not have national maritime borders like territorial seas or EEZs. In other words, the governance of marine resources, especially the governance of the high seas, should not just be regarded as an extension of land-based state borders into the ocean (Hung & Lien, 2022). Nonetheless, a question remains: what is unique about high seas territory?

More attention should be paid to the difference between the high seas and other oceanic spaces, such as territorial seas and EEZs. Indeed, scholars are increasingly attentive to the need to view modes of high seas territoriality as a distinct *form* of state intervention. For example, by studying the territorialisation of Areas Beyond National Jurisdiction (ABNJ),Lambach (2021) believes that the territoriality of the high seas goes beyond the scope of enclosure behaviours. As Lambach (ibid.) points out, the enclosure literature has focused primarily on changes within national waters and thus has hardly considered the territoriality of the high seas as a distinct concept. Yet modes of enclosure—albeit different to those in territorial waters or the EEZ—do happen in the high seas.

Therefore, following Lambach's argument, I argue that a first step in elucidating the territorial distinctiveness of the high seas is to examine the role of states *beyond* their bordering actions and capacities *within* national waters. For example, compared with territorial seas and EEZs, which have relatively clear ocean borders with their legality regarding the governance rights of the nation-state, the territoriality of the high seas requires countries to adopt different territorial strategies, which are not strict and fixed with maritime boundaries for ocean containment (Peters, 2020a, 2020b). This includes establishing bodies such as Regional Fisheries Management Organizations (RFMOs), which will be discussed shortly. Indeed, I take high seas territoriality as a *bordered-out territoriality*, through which the state performs its territorial power outside of its bordered sovereign rights in terrestrial terrains, territorial seas and EEZs. Furthermore, the uniqueness of state territoriality on the high seas also arises from the participation of non-state actors (Petersson et al., 2019) and the ecological materiality of the high seas (Teo, 2023). This means that the 'bordered-out' territoriality of the state beyond the legal space of its sovereign rights has forged a new mode of territoriality and control, which is constituted through the networks among the state, non-state actors and the nonhuman agency of the ocean.

Again, I take fisheries, especially distant-water fisheries operating in the high seas, as a useful example for demonstrating 'bordered-out' modes of territory-making, through the intersection of state power, practices of non-state actors (like capitalist enterprises or particularly as this section shows, fisheries scientists) and the nonhuman agency of the ocean (such as fish migration). Similar to ocean spaces closer to land (the territorial sea and EEZ), the depletion of high seas resources such as migratory fish has been recognised as a problem worldwide (Dupont & Baker, 2014; Fabinyi, 2020; Havice, 2018; Kim, 2019; Marshall, 2004). Therefore, 'effective' governance of fish stocks of the migrant species has become one of the political and economic strategies for managing total allowable catches in the high seas. Here the governance of marine resources in the high seas requires the development of specific fishery sciences. More specifically, the production of scientific knowledge for managing fish resources is often associated with two purposes. One is to obtain precise quantities of the existing fish stock in order to allocate future allowable fishing quotas. The second is to monitor changes in the overall population of target fish to achieve effective management of fishing activities through the restriction, for example, of further fishing. Thus, it is common today for both state and non-state actors or institutions, including governments, environmental NGOs and fishing companies, to rely on the practices of science-driven governance of fish stocks to territorially manage areas of ocean beyond national jurisdiction (Dupont et al., 2020).

The pursuit of scientific governance over the fish stocks, whether it is the accurate quantity of catch or the precise location of fishing areas, becomes a legitimising mechanism for high seas governance between different states. In other words, scientific data on fish stocks, with so-called scientific techniques and counting methods, are labelled by governments, environmental NGOs and fishing companies as tools to produce 'neutral' data for the 'equitable' allocation and governance of the ocean resources (Polacheck, 2012). However, science-driven governance of ocean resources and its scientific claims enable the state and non-state actors to expand control of the high seas in the name of scientific accuracy to achieve equity, justice and sustainability (Boucquey et al., 2019; Havice et al., 2022). Paradoxically, so-called scientific accuracy also becomes political rhetoric when used as a tool in negotiations for high seas governance. While it is difficult to achieve complete scientific accuracy in any modern data collection, the pursuit of more precise numbers or scientific results is a never-ending task. This does not negate the scientific data or results produced by fisheries scientists. Rather, it calls for a more critical perspective to investigate the *apolitical* discourse of ocean science embedded in the data produced and its use in inter-state governance over the ocean space (Boucquey, 2020), especially in the relationship between the distant-water fishery and the high seas.

One dimension of the shift from an apolitical to a political perspective in science-driven ocean governance over the high seas is to address the statecraft of territory through the generation of fish population profiles. I therefore argue that data-production-based ocean science, particularly fisheries science, has become a mechanism for extending high seas governance among states in the name of resource sustainability. That is, through the production of so-called precise figures, maritime space can be calculated, defined and controlled in negotiations, and even in conflicts, between different states and between state and non-state actors. Yet no scientific analysis can claim absolute certainty in the data for the governance of any marine resource. In other words, behind the scientific data collected by fisheries scientists are long-standing scientific uncertainties. Uncertainties create ambiguities, prompting negotiations between different states and between state and non-state actors. Therefore, the ambiguity underlying scientific uncertainty is also crucial for establishing new regimes of RFMO, which work to territorialise the high seas under the remit of oceanic resource governance. The margin between scientific ambiguity and accuracy, then, becomes an *operational strategy* for states to extend their territorialities to the high seas within the framework of RFMOs. In other words, RFMOs become an emerging regime for a collective inter-state governance over the high seas outside the state-bordered sovereignty in territorial seas and EEZs.

To sum up for this section, under the framework of RFMOs, fisheries science thus both invites and legitimises state political action towards the territorialisation of high seas. On the one hand, due to the uncertainty of scientific data about fish stocks, science has opened the door for the state to step in and work with fisheries scientists to defend the best interests of the state. On the other hand, fisheries science paradoxically also appears to depoliticise the political behaviour of states, intervening in the territorialisation of RFMOs in the form of scientific management of fish stocks. Having said that, fisheries science not only produces data for the analysis of fish populations, but also produces RFMO territories where the states are actively involved in constructing the governance of the high seas.

Overlapped Territoriality: Maritime Grey Zones

Finally, this chapter will discuss ocean borders around so-called maritime grey zones. In particular, I highlight the situations concerning disputed maritime spaces and their relationship with state sovereign rights over a variety of marine resources, including fish. For example, disputed islands all over the world—the Chagos Archipelago in the Indian Ocean, the Falkland Islands/Islas Malvinas in the Atlantic Ocean, the Islets of the South China Sea, the Dokdo/Takeshima Island, the Senkaku/Diaoyu Island in the western Pacific, to name but a few—have long been related to the geopolitical construction of state territorial contestations over the ocean. Meanwhile, recent discussions regarding the emergence of a new geopolitical divide, or 'new cold war', have furthermore put some disputed islands into a new phase of territorial conflict over the ocean. For example, China has strengthened its sovereign claims over the South China Sea by building up the military infrastructures on the islets and even the coral reefs (Hung & Lien, 2022). As a result, China's actions on the islets of the South China Sea have increased tension concerning the territorial disputes among China, Taiwan, Vietnam, the Philippines and Malaysia.

The recent development of the 'maritime grey zone' concept, indeed, exemplifies a complicated relationship between sovereignty and territory. Maritime borders and the extensions of a state's sovereign claims 'do not necessarily index an extension of sovereign territory' (Billé, 2020: 2). In other words, discussions of ocean-going state power are situated in a zone of political complexity where territory and sovereignty are not necessarily coexistent. This trend has pushed border studies to rethink issues of ocean borders without necessarily assuming that sovereignty implies a bounded state territory. Maritime grey zones therefore require an alternative analysis that is attuned to the ways in which such ocean border practices dynamically disrupt, transform and reorganise relations between states, sovereignty, territories and borders.

Innovative boundary practices are crucial in this regard to maritime grey zones. In the South China Sea, again for example, China has militarised its domestic fishing fleet as a strategy to deploy its power and legitimise its exercise of sovereignty within its disputed 'nine-dash line' concerning the ocean border claims (Kennedy, 2018; Patalano, 2018; Zhang, 2016). In addition to treating its fishing fleet as a mobile border *infrastructure*, China also promotes 'patriotic tourism' in the South China Sea (Huang & Suliman, 2020), with tourists emotionally embodying China's claims to this ocean borders (Rowen, 2018). Patriotic tourism consists of performative practices and strategic actions oriented towards a 'spatial politics of affects' (Thrift, 2004) that seek to enhance and legitimise nationalist feelings about maritime territories. Additionally, Chinese dredgers in the region have become a geopolitical issue between China, Taiwan and other countries surrounding the South China Sea. At the same time, the Taiwanese government has designated the adjacent offshore area of the Taiwan Strait as a wind farm area, aiming to develop Taiwan's wind power capacity. As a result, Taiwan Strait, as a maritime grey zone under the geopolitical tension between China and Taiwan, has re-bordered by the Taiwan state and then redefined as territories for the development of green energy (Hung, 2020). In other words, the state employs strategies such as new infrastructures and affective embodiments to territorialise the maritime grey zones.

Specifically, maritime grey zones are now often conceptualised as a state's strategy, which is nevertheless deliberately calculated to remain below a level that would trigger an armed response to the disputed ocean space (Harold et al. 2017). The disputed Diaoyu/Senkaku Island among Japan, Taiwan and China has been an example (Harold et al. 2017; Wirth, 2016) to address the issue. A 2012 fisheries agreement between Taiwan and Japan designates a 74,000 square kilometre bordered area around the Diaoyu/Senkaku Islands. In this maritime grey zone, Taiwan and Japan share the rights to fish, patrol and enforce laws (Yeh et al., 2015), but they deny these rights to other countries, including China. More importantly, neither Taiwan nor Japan asserts or deploys sovereign power in full accordance with traditional definitions of land sovereignty and territory. In other words, the fisheries agreement between Taiwan and Japan displays a cooperative bordering force in claiming both states' sovereign rights over the fishing resources without directly advancing the conflict regarding the sovereignty over the disputed Diaoyu/Senkaku Island. The bordered grey zone based on the fisheries agreement between Taiwan and Japan creates overlapped territorialities for fishing with elastic sovereign rights between the two states.

In sum, the overlapped territorialities in the maritime grey zones, as illustrated here in the case of fishing rights, demonstrate the elasticity of the state governance through the bordering of oceanic spaces. This scholarship has provided a critical way to challenge the static and very often dematerialised thinking of borders that derive from the terrestrial concepts of state territoriality (Hung & Lien, 2022). Next, I conclude the chapter by briefly addressing the reconceptualisation of ocean borders within mobile practices and the dynamic materiality of the ocean.

Ocean Border Reconsidered: Fisheries and Mobile State Power

Recent scholarship regarding the ocean has highlighted the material differences between *sea*scape and *land*scape (Childs, 2019; Duara, 2021; Steinberg & Peters, 2015). However, this does not necessarily mean that the materiality of *sea*scape has been a reified distinctiveness as an absolute opposition to the *land*scape (Lambach, 2021). As discussed in previous sections, bordering forces have been prevalent in forging *both* terrestrial and oceanic spaces. As such, ocean borders based on the state sovereign rights have been a form of power in managing and governing the ocean, as seen in the earlier sections of this chapter (Hung & Lien, 2022). Nevertheless, facing the challenge of acknowledging the significance of oceanic materiality without reifying another binary opposition between marine and terrestrial spaces, scholars in critical ocean studies do need to reconsider forms of resource governance over the ocean (Lambach, 2021; Peters, 2020a, 2020b). These discussions have been associated with the rethinking state territoriality and practices of sovereign rights over the oceanic space in their different guises—territorial sea spaces, EEZs and the high seas (Havice, 2018; Hung & Lien, 2022).

While the political technology of state territoriality (Elden, 2010) and the associated construction of oceanic space never stops (Lambach, 2021), the ways for deploying the power of the state have transformed. It is this point with which the chapter concludes. Mobile state governance has been critical in investigating the changing relationship between the state sovereign rights and the maritime space (Hung & Lien, 2022). In particular, in comparison with the way of controlling land-based territory, as the state enacts nuanced responses to multiple ocean materialities, it also exercises its power in an increasingly mobile fashion. In other words, when governing fluid waters and migrating fish, for example, the state must itself not only reconfigure the legality for oceanic governance (Braverman & Johnson, 2020), but also make governance more mobile in order to more effectively use and control the ocean (Peters, 2014). As a result, the laws of the sea, in particular the regulations on the fishing activities and the related actions of moving boats, therefore, have been one of the manifestations of a mobile state governance.

With regard to the territorialisation of the ocean, the idea of mobile state governance, however, does not refer to an unlimited reaching out of state power in regulating the moving fishing activities. Whether by national legislation or international agreement, the boundaries of the ocean that delineate the area of ocean fishing law continue to restrict mobile state governance and its practices of sovereign rights. However, the laws of the sea move with an inevitable attentiveness to the nonhuman materialities of the ocean, just as the moving boats and fish challenge the modes of state governance and the sovereign rights.

For future critical studies concerning ocean borders, one important dimension is the tension between the state's struggle over oceanic materialities, be they the tidal waves or the migrating fish, and the state's action in constantly 'projecting its will to dominate, know, and impose material marks over the fluidity and mobility of the sea' (de Lucia, 2023: 121). Nevertheless, projecting state sovereign rights over the ocean does not equal the effective and practical control over ocean space. In other words, the imposition of material markers still needs a mechanism to operate, perform and realise the effectiveness of state sovereign rights. Governance of fisheries resources, or more broadly the effective practices of managing resources per se of a sea, I argue, is one of the critical issues for future ocean studies to look into as the state expands of sovereign power to the vast breadths and depths of maritime space.

Acknowledgements Completion of this chapter is supported by the National Science and Technology Council in Taiwan (110-2410-H-002-186; 111-2410-H-002-099-MY3) and Taiwan Social Resilience Research Center (Grant no. 112L900303) from the Higher Education Sprout Project by the Ministry of Education in Taiwan.

References

Alexander, L. M. (1968). Geography and the law of the sea. *Annals of the Association of American Geographers, 58*(1), 177–197.

Barbesgaard, M. (2018). Blue growth: Savior or ocean grabbing? *The Journal of Peasant Studies, 45*(1), 130–149.

Benjaminsen, T. A., Buhaug, H., McConnell, F., & Sharp, J. P. (2017). Political geography and the environment. *Political Geography, 100*(56), A1–A2.

Bennett, N. J. (2019). In political seas: Engaging with political ecology in the ocean and coastal environment. *Coastal Management, 47*(1), 67–87.

Bennett, E. M., Cramer, W., Begossi, A., Cundill, G., Díaz, S., Egoh, B. N., Geijzendorffer, I. R., Krug, C. B., Lavorel, S., Lazos, E., Lebel, L., Martín-López, B., Meyfroidt, P., Mooney, H. A., Nel, J. L., Pascual, U., Payet, K., Harguindeguy, N. P., Peterson, G. D., ... & Woodward, G. (2015). Linking biodiversity, ecosystem services, and human well-being: Three challenges for designing research for sustainability. *Current Opinion in Environmental Sustainability, 14*, 76–85.

Billé, F. (2020). *Voluminous states: Sovereignty, materiality, and the territorial imagination.* Duke University Press.

Bocarejo, D., & Ojeda, D. (2016). Violence and conservation: Beyond unintended consequences and unfortunate coincidences. *Geoforum, 69*, 176–183.

Boucquey, N. (2020). The 'nature' of fisheries governance: Narratives of environment, politics, and power and their implications for changing seascapes. *Journal of Political Ecology, 27*(1), 169–189.

Boucquey, N., Martin, K. S., Fairbanks, L., Campbell, L. M., & Wise, S. (2019). Ocean data portals: Performing a new infrastructure for ocean governance. *Environment and Planning D: Society and Space, 37*(3), 484–503.

Braverman, I., & Johnson, E. (2020). *Blue legalities: The life and laws of the sea.* Duke University Press.

Bridge, G. (2014). Resource geographies II: The resource-state nexus. *Progress in Human Geography, 38*(1), 118–130.

Campbell, L. M. (2007). Local conservation practice and global discourse: A political ecology of sea turtle conservation. *Annals of the Association of American Geographers, 97*(2), 313–334.

Campling, L., & Colás, A. (2018). Capitalism and the sea: Sovereignty, territory and appropriation in the global ocean. *Environment and Planning D: Society and Space, 36*(4), 776–794.
Campling, L., & Havice, E. (2014). The problem of property in industrial fisheries. *The Journal of Peasant Studies, 41*(5), 707–727.
Carothers, C. (2011). Equity and access to fishing rights: Exploring the community quota program in the Gulf of Alaska. *Human Organization, 70*(3), 213–223.
Carothers, C., & Chambers, C. (2012). Fisheries privatization and the remaking of fishery systems. *Environment and Society, 3*(1), 39–59.
Carver, R., Childs, J., Steinberg, P. E., Mabon, L., Matsuda, H., Squire, R., McLellan, B., & Esteban, M. (2020). A critical social perspective on deep sea mining: Lessons from the emergent industry in Japan. *Ocean & Coastal Management, 193*, 1–10.
Childs, J. (2019). Greening the blue? Corporate strategies for legitimising deep sea mining. *Political Geography, 74*, 102060. https://doi.org/10.1016/j.polgeo.2019.102060
Couper, A. D. (1978). Marine resources and environment. *Progress in Human Geography, 2*(2), 296–308.
De Lucia, V. (2023). Anthropocentrism and international environmental law. In V. Capaux, F. Mégret, & U. Natarajan (Eds.), *The Routledge handbook of international law and anthropocentrism* (pp. 84–101). Routledge.
De Santo, E. M. (2020). Militarized marine protected areas in overseas territories: Conserving biodiversity, geopolitical positioning, and securing resources in the 21st century. *Ocean & Coastal Management, 184*, 105006. https://doi.org/10.1016/j.ocecoaman.2019.105006
De Santo, E. M. (2024). *Securitizing marine protected areas geopolitics, environmental justice, and science*. Routledge.
Dittmer, J., Moisio, S., Ingram, A., & Dodds, K. (2011). Have you heard the one about the disappearing ice? Recasting Arctic Geopolitics. *Political Geography, 30*(4), 202–214.
Dodds, K., & Benwell, M. C. (2010). More unfinished business: The Falklands/Malvinas, maritime claims, and the spectre of oil in the South Atlantic. *Environment and Planning D: Society and Space, 28*(4), 571–580.
Drewniak, M., Dalaklis, D., Kitada, M., Ölçer, A., & Ballini, F. (2018). Geopolitics of Arctic shipping: The state of icebreakers and future needs. *Polar Geography, 41*(2), 107–125.
Duara, P. (2021). Oceans as the paradigm of history. *Theory, Culture & Society, 38*(7–8), 143–166.

Duffy, R. (2014). Waging a war to save biodiversity: The rise of militarized conservation. *International Affairs, 90*(4), 819–834.

Dunlap, A., & Fairhead, J. (2014). The militarisation and marketisation of nature: An alternative lens to 'climate-conflict.' *Geopolitics, 19*(4), 937–961.

Dupont, A., & Baker, C. G. (2014). East Asia's maritime disputes: Fishing in troubled waters. *The Washington Quarterly, 37*(1), 79–98.

Dupont, C., Herpers, F., & Le Visage, C. (2020). *Recommendations for positive interactions between offshore wind farms and fisheries: Short background study.* European Commission, Executive Agency for Small and Medium-sized Enterprises, Publications Office. Retrieved May 24, 2024, from https://data.europa.eu/doi/10.2826/017304

Elden, S. (2010). Land, terrain, territory. *Progress in Human Geography, 34*, 799–817.

Ertör, I., & Hadjimichael, M. (2020). Blue degrowth and the politics of the sea: Rethinking the blue economy. *Sustainability Science, 15*, 1–10.

Fabinyi, M. (2020). Maritime disputes and seafood regimes: A broader perspective on fishing and the Philippines-China relationship. *Globalizations, 17*(1), 146–160.

Fabinyi, M., Foale, S., & Macintyre, M. (2015). Managing inequality or managing stocks? An ethnographic perspective on the governance of small-scale fisheries. *Fish and Fisheries, 16*(3), 471–485.

Fairbanks, L. (2019). Policy mobilities and the sociomateriality of US offshore aquaculture governance. *Environment and Planning C: Politics and Space, 37*(5), 849–867.

Fairbanks, L., Campbell, L. M., Boucquey, N., & St. Martin, K. (2018). Assembling enclosure: Reading marine spatial planning for alternatives. *Annals of the American Association of Geographers, 108*(1), 144–161.

Fawcett, L., Havice, E., & Zalik, A. (2022). Frontiers: Ocean epistemologies–privatise, democratise, decolonise. In K. Peters, J. Anderson, A. Davies, & P. E. Steinberg (Eds.), *The Routledge handbook of ocean space* (pp. 70–84). Routledge.

Figueroa, I., Satizábal, P., Saavedra-Díaz, L. M., Noriega-Narváez, G., & Velásquez-Mendoza, Y. (2023). Manifesto for the protection of the human rights of small-scale fishing communities in Colombia. *Jangwa Pana, 22*(3), 2–4.

Figueroa, I., Saavedra-Díaz, L. M., Satizábal, P., Noriega-Narváez, G., & Velásquez-Mendoza, Y. (2024). Justice in fishing territories: Human rights violations in artisanal fisheries analyzed by the Colombian Constitutional Court. *Journal of Political Ecology, 31*(1), https://doi.org/10.2458/jpe.6026

Foley, P., & Mather, C. (2019). Ocean grabbing, terraqueous territoriality and social development. *Territory, Politics, Governance, 7*(3), 297–315.

George, R. Y., & Wiebe, S. M. (2020). Fluid decolonial futures: Water as a life, ocean citizenship and seascape relationality. *New Political Science, 42*(4), 498–520.

Gray, N. J. (2018). Charted waters? Tracking the production of conservation territories on the high seas. *International Social Science Journal, 68*(229–230), 257–272.

Hadjimichael, M. (2022). (De)growth: The right to the sea. In K. Peters, J. Anderson, A. Davies, & P. E. Steinberg (Eds.), *The Routledge handbook of ocean space* (pp. 161–172). Routledge.

Harold, S. W., Nakagawa, Y., Fukuda, J., Davis, J. A., Kono, K., Cheng, D., & Suzuki, K. (2017). *The US Japan alliance and deterring gray zone coercion in the maritime, cyber, and space domains.* RAND Corporation. Retrieved May 24, 2024, from https://doi.org/10.7249/CF379

Hasty, W., & Peters, K. (2012). The ship in geography and the geographies of ships. *Geography Compass, 6*(11), 660–676.

Havice, E. (2018). Unsettled sovereignty and the sea: Mobilities and more-than-territorial configurations of state power. *Annals of the American Association of Geographers, 108*(5), 1280–1297.

Havice, E., & Zalik, A. (2018). Ocean frontiers: Epistemologies, jurisdictions, commodifications. *International Social Science Journal, 68*(229–230), 219–235.

Havice, E., Campbell, L., & Boustany, A. (2022). New data technologies and the politics of scale in environmental management: Tracking Atlantic bluefin tuna. *Annals of the American Association of Geographers, 112*(8), 2174–2194.

Huang, Y., & Suliman, S. (2020). Geopolitics, (re)territorialisation, and China's patriotic tourism in the South China Sea. *Geopolitics, 98*, 102669. https://doi.org/10.1016/j.polgeo.2022.102669

Hung, P. Y. (2020). Placing green energy in the sea: Offshore wind farms, dolphins, oysters, and the territorial politics of the intertidal zone in Taiwan. *Annals of the American Association of Geographers, 110*(1), 56–77.

Hung, P. Y., & Lien, Y. H. (2022). Maritime borders: A reconsideration of state power and territorialities over the ocean. *Progress in Human Geography, 46*(3), 870–889.

Kamat, V. R. (2018). Dispossession and disenchantment: The micropolitics of marine conservation in southeastern Tanzania. *Marine Policy, 88*, 261–268.

Kennedy, C. (2018). The struggle for blue territory: Chinese maritime militia grey-zone operations. *The RUSI Journal, 163*(5), 8–19.

Kim, H. J. (2019). South Korea's use of force against Chinese illegal fishing in the course of law enforcement in the Yellow Sea. *Marine Policy, 99*, 148–156.

Knott, C., & Neis, B. (2017). Privatization, financialization and ocean grabbing in New Brunswick herring fisheries and salmon aquaculture. *Marine Policy, 80*, 10–18.

Lambach, D. (2021). The functional territorialization of the high seas. *Marine Policy, 130*, 104579. https://doi.org/10.1016/J.MARPOL.2021.104579

Marshall, J. (2004). Defining maritime boundaries: 'The murky hand of history's oversight' in the Gulf of Maine. *Canadian Geographer/le Géographe Canadien, 48*(3), 266–286.

Mulazzani, L., & Malorgio, G. (2017). Blue growth and ecosystem services. *Marine Policy, 85*, 17–24.

Muralidharan, R., & Rai, N. D. (2020). Violent maritime spaces: Conservation and security in gulf of Mannar Marine National Park, India. *Political Geography, 80*, 102160. https://doi.org/10.1016/j.polgeo.2020.102160

Neumann, R. P. (2004). Moral and discursive geographies in the war for biodiversity in Africa. *Political Geography, 23*(7), 813–837.

Østhagen, A. (2020). Maritime boundary disputes: What are they and why do they matter? *Marine Policy, 120*, 104118. https://doi.org/10.1016/j.marpol.2020.104118

Patalano, A. (2018). When strategy is 'hybrid' and not 'grey': Reviewing Chinese military and constabulary coercion at sea. *The Pacific Review, 31*(6), 811–839.

Peluso, N. L. (1993). Coercing conservation? The politics of state resource control. *Global Environmental Change, 3*(2), 199–217.

Peluso, N. L., & Vandergeest, P. (2011). Political ecologies of war and forests: Counterinsurgencies and the making of national natures. *Annals of the Association of American Geographers, 101*(3), 587–608.

Peluso, N. L., & Watts, M. (2001). *Violent environments*. Cornell University Press.

Peters, K. (2014). Tracking (im)mobilities at sea: Ships, boats and surveillance strategies. *Mobilities, 9*(3), 414–431.

Peters, K. (2020a). Deep routeing and the making of 'Maritime Motorways': Beyond surficial geographies of connection for governing global shipping. *Geopolitics, 25*(1), 43–64.

Peters, K. (2020b). The territories of governance: Unpacking the ontologies and geophilosophies of fixed to flexible ocean management, and beyond. *Philosophical Transactions of the Royal Society B, 375*(1814), 20190458. https://doi.org/10.1098/rstb.2019.0458

Peters, K., & Steinberg, P. E. (2019). The ocean in excess: Towards a more-than-wet ontology. *Dialogues in Human Geography, 9*(3), 293–307.

Peters, K., Steinberg, P. E., & Stratford, E. (Eds.). (2018). *Territory beyond terra*. Rowman & Littlefield.

Petersson, M. T., Dellmuth, L. M., Merrie, A., & Österblom, H. (2019). Patterns and trends in non-state actor participation in regional fisheries management organizations. *Marine Policy, 104*, 146–156.

Pinkerton, E., & Davis, R. (2015). Neoliberalism and the politics of enclosure in North American small-scale fisheries. *Marine Policy, 61*, 303–312.

Polacheck, T. (2012). Assessment of IUU fishing for Southern Bluefin Tuna. *Marine Policy, 36*(5), 1150–1165.

Rasmussen, M. B., & Lund, C. (2018). Reconfiguring Frontier spaces: The territorialization of resource control. *World Development, 101*, 388–399.

Raycraft, J. (2019). Circumscribing communities: Marine conservation and territorialization in southeastern Tanzania. *Geoforum, 100*, 128–143.

Rowen, I. (2018). Tourism as a territorial strategy in the South China Sea. In K. Spangler (Ed.), *Enterprises, localities, people, and policy in the South China Sea* (pp. 61–74). Palgrave Macmillan.

Satizábal, P., & Dressler, W. H. (2019). Geographies of the sea: Negotiating human–fish interactions in the waterscapes of Colombia's Pacific Coast. *Annals of the American Association of Geographers, 109*(6), 1865–1884.

Steinberg, P. E., & Peters, K. (2015). Wet ontologies, fluid spaces: Giving depth to volume through oceanic thinking. *Environment and Planning D: Society and Space, 33*(2), 247–264.

Stern, M. J. (2008). Coercion, voluntary compliance and protest: The role of trust and legitimacy in combating local opposition to protected areas. *Environmental Conservation, 35*(3), 200–210.

Tafon, R. V. (2018). Taking power to sea: Towards a post-structuralist discourse theoretical critique of marine spatial planning. *Environment and Planning C: Politics and Space, 36*(2), 258–273.

Teo, J. (2023). Ocean materialities and the ontologies of ocean conservation in areas beyond national jurisdiction: The Costa Rica Thermal Dome. *Marine Policy, 153*, 105646. https://doi.org/10.1016/j.marpol.2023.105646

Thrift, N. (2004). Intensities of feeling: Towards a spatial politics of affect. *Geografiska Annaler: Series B, Human Geography, 86*(1), 57–78.

TNI Agrarian Justice Programme, Masifundise and Afrika Kontakt. (2014). *The global ocean grab: A primer*. Retrieved March 17, 2020 from https://www.tni.org/files/download/the_global_ocean_grab.pdf
Vandergeest, P., & Peluso, N. L. (1995). Territorialization and state power in Thailand. *Theory and Society, 24*(3), 385–426.
Voyer, M., Quirk, G., McIlgorm, A., & Azmi, K. (2018). Shades of blue: What do competing interpretations of the Blue Economy mean for oceans governance? *Journal of Environmental Policy & Planning, 20*(5), 595–616.
Winder, G. M., & Le Heron, R. (2017). Assembling a Blue Economy moment? Geographic engagement with globalizing biological-economic relations in multi-use marine environments. *Dialogues in Human Geography, 7*(1), 3–26.
Wirth, C. (2016). Securing the seas, securing the state: Hope, danger and the politics of order in the Asia-Pacific. *Political Geography, 53*, 76–85.
Ybarra, M. (2016). "Blind passes" and the production of green security through violence on the Guatemalan border. *Geoforum, 69*, 194–206.
Yeh, Y. H., Tseng, H. S., Su, D. T., & Ou, C. H. (2015). Taiwan and Japan: A complex fisheries relationship. *Marine Policy, 51*, 293–301.
Zalik, A. (2018). Mining the seabed, enclosing the Area: Ocean grabbing, proprietary knowledge and the geopolitics of the extractive frontier beyond national jurisdiction. *International Social Science Journal, 68*(229–230), 343–359.
Zaucha, J., & Jay, S. (2022). The extension of marine spatial planning to the management of the world ocean, especially areas beyond national jurisdiction. *Marine Policy, 144*, 105218. https://doi.org/10.1016/j.marpol.2022.105218
Zhang, H. (2016). Chinese fishermen in disputed waters: Not quite a "people's war." *Marine Policy, 68*, 65–73.

Contested Borders and Resolution in Planning Shared Marine Waters

Joseph Onwona (Kofi) Ansong

Introduction

The years from 2020 have been described as 'super years' for the ocean by the global ocean community due to increased political attention towards sustainable ocean use, protection, restoration, and safeguarding for future generations. Amidst these efforts, there is growing global concern regarding the deteriorating marine environment, ocean oxygen decline, the loss of marine biodiversity and habitats, marine pollution, impact of climate change on coastal communities, as well as the disjointed international efforts in addressing these global marine problems. These marine problems are exacerbated in contested marine areas where questions of jurisdiction are unclear and subject to negotiation (this equates to 55% of maritime boundaries). In this chapter, borders and boundaries are

J. O. (Kofi) Ansong (✉)
Department of Geography and Planning, University of Liverpool, Liverpool, UK
e-mail: j.ansong@liverpool.ac.uk

Howell Marine Consulting, Low Hauxley, England

K. Peters and J. Turner (eds.), *Ocean Governance (Beyond) Borders*,
https://doi.org/10.1007/978-3-031-71322-4_5

used interchangeably. Here, a *border* denotes an internationally recognised line on a map separating two sovereign states (Ritchie et al., 2019). Likewise, a *boundary* line signifies the extent and limits of jurisdictional powers and the allocation of natural resources (Forbes, 2001). In contested marine areas, there are issues of resource conflicts, undermined marine policies, and contestation over international agreements (UN, 2021; Kuempel et al., 2019; Østhagen, 2019). These challenges highlight the urgent need for radical planning approaches which go beyond border limitations to manage the mobile, fluid, dynamic, three-dimensional and transboundary realm of the sea in Marine Spatial Planning (MSP) (see also Couling, this volume; McAteer, this volume).

Recent literature has critiqued existing ocean management and MSP approaches as they have the tendency to create bordered zones within which to control human activities and marine resources. For example, Peters (2020) refers to current ocean management approaches as 'territorial traps' or 'carceral' seas and argues for de-territorialising ocean governance by promoting more flexible dynamic approaches. Faludi (2019) notes that MSP is burdened with territoriality, which is the building block of the political order in a nation-state system and creates a delusion of territorial sovereignty. Ritchie et al. (2019) suggest that assortment of transboundary marine-related institutions creates contradictory arrangements of 'boundary-ness' for the same marine border.

There are different interpretations of marine boundaries and conceptual propositions about their consideration in MSP, including MSP assemblages[1] (Fairbanks et al., 2018), soft spaces[2] (Faludi, 2013), lively spaces[3] (Jay, 2018), and dynamic ocean management[4] (Maxwell et al., 2015). Yet, literature that examines how marine spatial plans are considering contested borders and related resource conflict is limited. This

[1] This refers to the constellation of actors and relations involved in an indeterminate process and planning of marine space.

[2] This approach considers associational relationships which stretch across a range of geographies, and spaces through the use of visualisation networks and flows through the use of 'scenarios' and 'fuzzy maps'.

[3] Soft spaces that consider ecosystems and human interaction.

[4] Mapping and evidence which consider the rapid change in space and time in response to the shifting nature of marine species, habitats, and users.

chapter seeks to fill such gaps by asking the question: *how are jurisdictional issues around borders considered in MSP and can MSP address spatial conflicts in contested marine areas?*

A multiple case study approach is applied to understand contested boundaries and various drivers in Pomeranian Bay and the Island of Ireland (IOI). The two cases were selected due to the mature state of MSP in Ireland, the UK, and Germany (and its border with Poland). A brief history of maritime borders is discussed in the section to follow. I then go on to review existing MSP documents to understand how they have considered border issues and related conflicts in a detailed examination of the two cases mentioned. I conclude by drawing comparison between the two cases and highlight key findings to advance planning contested marine waters.

Contested Maritime Borders and Ownership: A Historical Issue

The history of maritime borders and delimitation emerged around the fifteenth century when Western European countries began to pursue colonialism which was largely attributed by a substantial trade with Africa and Asia via the Atlantic (Acemoglu et al., 2005). This trend instigated debates concerning ownership of the oceans and resource rights dominated by two contradictory principles: one relating to freedom (*Mare Liberum)* and the other relating to sovereignty and the closed sea (*Mare Clausum*). The idea of the sea being closed was mainly propounded by jurists including Welwod, de Freitas, and Selden. In contrast, for *Mare Liberum*, Hugo Grotius argued that the sea areas were open to all and, as such, nobody had the right to deny others access to them (Maier, 2016). Steinberg (1999) argues that these two competing ways of thinking about the ocean rested on notions of territorialisation and de-territorialisation, of ocean space in which state ownership could be exercised, and ocean space where it could not. Ultimately, this demarcation set in motion broader questions of oceanic bordering. After World War II, nation states signalled a desire to move away from having rights solely in a narrow band along their coast (the territorial sea) to more wide-ranging powers to control and manage their mineral resources in adjoining waters (Alexander, 1986). This resulted in the United Nations Convention on the Law of the Sea (UNCLOS), which came with the establishment of Exclusive Economic Zones (of up to 200 nautical miles) and a legislative framework for delimiting maritime zones. The Convention was adopted and signed in 1982. The inception of UNCLOS and maritime delimitation was in tandem with demarcating sea spaces like land, as a functional tool that society was/is comfortable with (Fairbanks et al., 2018; Lambach, 2021; Peters, 2020). Maritime delimitation introduced a division or bifurcation of some sort into the world (Nail, 2016). Borders created direct and indirect functions as they can simultaneously enable or disable, separate and connect, serve as barriers or bridges between marine ecosystems, users, coastal communities, and governance institutions (O'Dowd, 2010).

The question that springs up here is *what issues arise when borders are enacted at sea and become contested?* The next sections discuss the history of the contested borders and jurisdictional issues in the Pomeranian Bay and on the IOI in practice. Both cases share similar context such as a contested border due to drivers such as colonialism and the redefinition of country borders, marine use spatial conflicts and increasing demands. I use these as a way to examine and argue how the effectiveness of MSP in contested areas remains 'a planners dream' that is not achievable due to geopolitical problems and historical legacy disputes that emanate from border issues. There is some opportunity here to 'think outside the box' in how MSP is operationalised across borders to deliver sustainable outcomes.

Redefinition of Borders in the Pomeranian Bay

The Pomeranian Bay is situated between Adlergrund and Oder Estuary, between Mecklenburg-Vorpommern (Germany) and West Pomerania region (Poland) (see Fig. 1). The lower river Oder, including the city and ports of Szczecin and Świnoujście (later called the Oder–Neisse line), was part of the Piast dynasty and formed part of Poland's western border from the tenth until the thirteenth century. In the following years, the boundaries of Pomerania became unclear and changed due to different ownership by countries such as Sweden, Denmark, and Prussia (Focus, 2013). During World War I, the Russian Empire proposed restoring the border, in the belief that it would provide protection against Germany. After Nazis gained power, the German territory to the east of the line was militarised by Germany in case of future war, leading to the spread of German people, culture, and language in Poland.

Before World War II, Poland's western border with Germany was fixed under the terms of the Treaty of Versailles of 1919 (Boemeke et al., 1998). The terms followed the historic border between the Holy Roman Empire and Greater Poland but were adjusted to reflect the ethnic compositions of small areas near the traditional provincial borders. This left Germany divided into two portions (Germany and East Prussia) by the Polish Corridor (Pomerania region) and the independent Free City

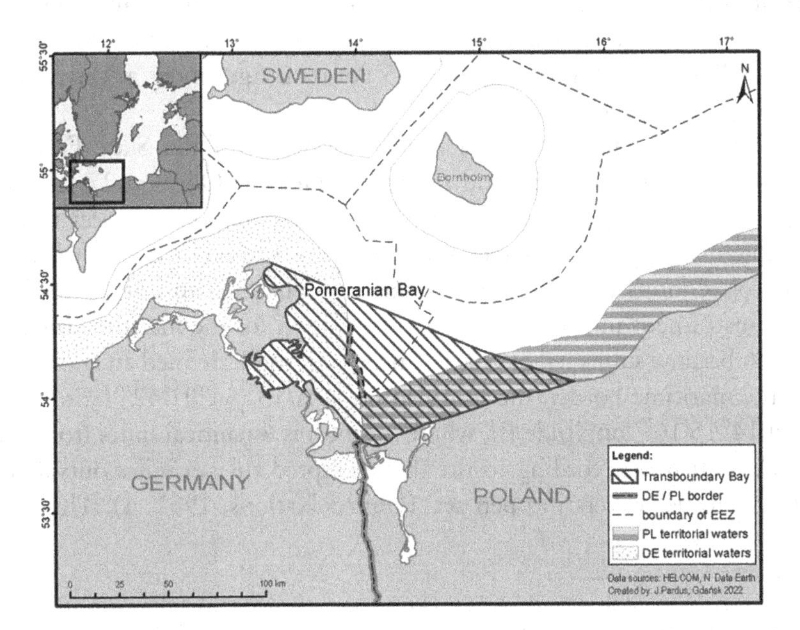

Fig. 1 The Pomeranian Bay and unclear maritime border at the northern approach to seaports in Poland highlighted in deep dashed lines. Source: Ansong et al. (2023: 2455), reproduced under the Creative Commons Attribution License [http://creativecommons.org/licenses/by/4.0/]

of Danzig (Gdansk). The Free City of Danzig had an overwhelmingly German population (96.7%) but was split from Germany to help secure Poland's strategic access to the Baltic Sea.

After World War II, the German-Polish Border was negotiated at the Potsdam Conference in August 1945 by the United States of America, UK, and Union of Soviet Socialist Republics (USSR). The German territories east of the Oder and Neisse River, including the ports of Szczecin and Świnoujście, were formally placed under Polish administrative control. Critically, the Treaty did not consider the demarcation of the territorial *waters* in the Pomeranian Bay, which had implications for the access routes to the Szczecin and Świnoujście ports. In 1945, a Polish-Soviet border commission was established but it did not feel competent to designate the border along the three miles of the Polish territorial sea. The unclear border in the Pomeranian Bay was repeated in the Zgorzelec Treaty signed on 6 July 1950 between the Polish Republic and German Democratic Republic (East Germany).[5] The Treaty recognised the Oder–Neisse line implemented by the 1945 Potsdam Agreement as the border between the two states.

The instrument confirming the Demarcation of the State Frontier between Germany and Poland signed in Frankfurt Oder on 27 January 1951 recognised the *special maritime delimitation circumstances* in the Pomeranian Bay. It was agreed that the border should be exempted from the conventions of UNCLOS, in this case, the median-line principle.[6] The instrument introduced a special regulation for the maritime border, which became subjected to bilateral negotiations: 'It defined an end point of the maritime border, the so-called point A/13 (54°01′42″ latitude N and 14°15′16″ longitude E), whose location is 6 nautical miles from the seashore that is according to the then accepted rules 3 miles outside of the territorial waters in open sea' (United Nations, 1951: 1). This was

[5] This treaty was not recognised by West Germany as a legitimate international treaty as the Soviet Union retained the rights to East Germany (similar to the rights that the Western Allies had in relation to West Germany under the Bonn and Paris Conventions). A final peace treaty with Germany (The Treaty of 1990) defined the final settlement regarding Germany after the Soviet Union granted the GDR independence.

[6] A nation's maritime boundaries should conform to a median-line equidistant from the shores of neighbouring nation-states

noted as unfavourable for Polish economic interest as the borderline east of the axis of the access route to the ports of Szczecin and Świnoujście left a part of the access route and one of the anchorages out of Polish territorial waters. The unusual location of *the point A/13* was a result of the alterations of the land borderline. The internal waters around the lake Wolgast were incorporated into Polish territory, so the German side received compensation of their surface in the area stretching from the inland water and the shoreline of the Pomeranian Bay. This exchange resulted in a deviation of the land frontier of which the marine border was an extension. Germany laid claim to the area based on the provisions in this agreement. However, the access route would then be on East Germany's territory and sea navigation along other, not deepened, routes in the Pomeranian Bay would be impossible for larger vessels.

The 1970 German Polish Normalisation Agreement signed in Warsaw established and normalised the relationship between West Germany and Poland. The Agreement repeated the same border description as the previous agreements in accordance with the Potsdam Agreement. The maritime border conflict gained prominence again in the second half of the 1980s when the two states of the socialist camp: Polish People's Republic (PPR) and the German Democratic Republic (GDR) stood against each other (Jackowska, 2008), and a unilateral decision was made by the GDR government on 20 December 1984, to extend their territorial sea up to 12 nautical miles. This affected shipping and port infrastructure elements (approach tracks and anchorages) that were necessary for the efficient operation of the Szczecin-Świnoujście Ports. The Polish side did not accept the actions of the GDR government to delimit the area as part of their territorial sea and created a disputed area in the Pomeranian Bay. This led to a series of conflict events between 1985 and 1989. In early 1985, GDR Navy ships approached vessels waiting to enter the Świnoujście port demanding that they leave the territorial waters of the GDR. In 1986, there were cases of ramming Polish boats and yachts by the GDR (Jackowska, 2008). Approximately, 180 of these incidents were reported and described by the Polish as having features of piracy and a threat to the safety of small boats (Ślepowroński, 2005).

On 22 May 1989, an Agreement was signed in Berlin by the Foreign Ministers to demarcate the borderline of the territorial sea, continental shelf, and fishery zones. This is alleged to be based on a proposed second compromise that left the entire access route to the Szczecin-Świnoujście ports, as well as the anchorage, located within the Polish territorial waters or in open sea. Poland refers to the 'Agreement between the PPR and the GDR on the Delimitation of the Sea Areas in the Oder Bay' on 22 May 1989 to claim ownership of the contested area. According to provisions of the above-mentioned agreement, the area of the approach channel to the ports of Szczecin and Świnoujście, as well as anchorage No. 3, lies in the Polish territorial sea or in the open sea. Article 5 of the Treaty states that:

> The entire extension of the North Approach to the ports of Szczecin and Swinoujscie and the anchorages are situated within the territorial sea of the Polish People's Republic or on the high seas.
> That section of the North Approach to the ports of Szczecin and Swinoujscie which is situated east of the outer limit of the territorial sea of the GDR as defined in Article 3 of the present Treaty and anchorage No. 3 shall not constitute continental shelf, fishery zone or potential exclusive economic zone of the GDR. (United Nations, 1989: 2)

Although the Treaty clarified that the anchorages do not constitute continental shelf and EEZ for the GDR, it was still not clear, as it stated that the northern approach could be situated in the high seas or territorial waters of Poland. It does not prevent and address which areas Germany can claim as territorial waters. The validity of the above-mentioned agreement and lack of clarity was repeated by the Treaty between the Republic of Poland and Federal Republic of Germany (after official unification of East and West Germany) on the confirmation of their existing border, concluded in Warsaw on 14 November 1990. Article 1 of the Treaty states that: 'The Contracting Parties reaffirm the frontier between them, whose course is defined in the Agreement between the Polish Republic and the GDR concerning the demarcation of the established and existing Polish-German State frontier of 6

July 1950 and agreements concluded with a view to implementing and supplementing ...' (United Nations, 1990: 1).

The Treaty repeats the agreement and provisions of the 1989 Agreement without any clarity about territorial seas demarcations. This lack of clarity about the ownership of the northern approach is still ongoing and was evidenced during the MSP process. During bilateral consultation between the Germany and Poland, the Polish Ministry of Infrastructure and the Polish Ministry of External Affairs notified the German Federal Ministry of the Interior for Building and Community that the 'contested area' is part of the Polish territorial sea and will not be depicted in the plan as a greyed-out area with unclear legal status. The request by the Germans to oppose the demarcation of the contested area for sea use and MSP purposes was not accepted by the Polish. The border issues in the Pomeranian Bay currently remain unclear amidst ongoing discussion between responsible ministries.

The unclear maritime border has contributed to competing interests on both sides of the border. On the Polish side, a key concern is to secure access to the ports of Szczecin and Świnoujście, which requires regular dredging activities; there are also plans to develop new port facilities and expand container facilities. On the German side, key interests are military use and Natura 2000 sites which protect bird species, harbour porpoise, and other important habitats. These uses make dredging and unrestricted shipping in Polish ports even more problematic. Beyond these interests, the disputed *Nord stream 1 and 2* gas pipelines are situated close to the contested area.

Contested Ownership of the Cross-Border Loughs: Island of Ireland

This case study was selected alongside the last as it presents border disputes on an island with two different governance systems as compared to the continental ecosystem in the Pomeranian Bay. The IOI, as a single biogeographical feature, comprises the Republic of Ireland (ROI) and Northern Ireland (NI), which is a devolved administration of the United Kingdom (UK) (see Fig. 2). Maritime boundaries on the island end in shared bays (Loughs) namely Lough Foyle and Carlingford Lough. The ownership of the Lough's seabed is disputed and there is no agreement on maritime border delimitation between ROI and NI.

Until the Middle Ages (circa 546 AD–1500 AD), fishing on the Loughs was controlled by religious Monasteries. In 1613, the Honourable Irish Society[7] took over the ownership and management of fishing rights on Lough Foyle from the Irish Bishops as part of overseeing the Plantation of Ulster. In the late 1600, the Royal Charter that established the county of Londonderry included all the waters of Lough Foyle up to the high-water mark on the Donegal side right up to Lifford within the county boundaries to protect salmon fishery rights (Symmons,

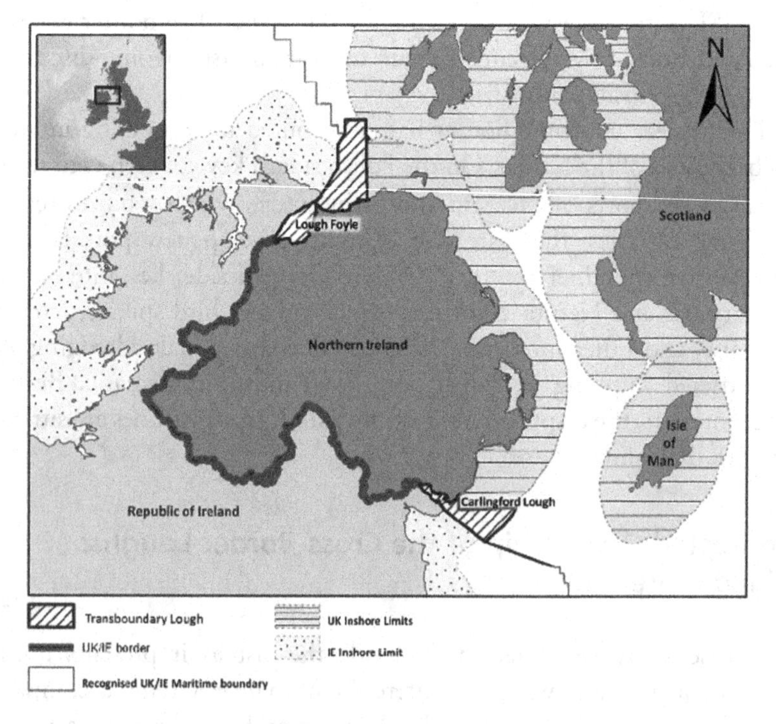

Fig. 2 Contested cross-border Loughs on the IOI. Produced by the author

[7] A consortium of trade associations and companies from London established to manage the Plantation of Ulster and colonise County Londonderry.

2009). Fishing rights in Lough Foyle were subsequently purchased by the Honourable Irish Society in 1704. The partition of Ireland in 1921 and establishment of NI (6 parliamentary counties[8]) governed by Britain resulted in ambiguity and debate about who owned fishing rights of Lough Foyle and the delimitation of territorial waters (Flannery et al., 2015). At the time of partition, it was envisaged that NI and ROI would both remain part of the UK (NIA, 2010).

The Government of Ireland Act of 1920 did not address the sea boundary issue in the Loughs. The creation of the Irish Free State, and consequently ROI, broke governance and political links between ROI and the UK, and dispute over Irish and British territorial waters took on a new significance. The ROI laid constitutional claim to all the waters around the IOI under the former Articles 2 and 3 of the Irish Constitution Act, 1937:

> Old Article 2: The national territory consists of the whole island of Ireland, its islands, and the territorial seas.
> Old Article 3: Pending the re-integration of the national territory, and without prejudice to the right of the Parliament and Government established by this Constitution to exercise jurisdiction over the whole of that territory, the laws enacted by that Parliament shall have the like area and extent of application as the laws of Saorstát Eireann [the Irish Free State] and the like extra-territorial effect. (Office of the Attorney General, 1937: 1)

The poaching of salmon fish during the 1930–1940s and the lack of regulatory body on both sides of the border led to the establishment of the Foyle Fisheries Acts in 1952 by both governments. These Acts led to the creation of a cross-border Foyle Fisheries Commission (FFC) in 1952 to control salmon poaching. The FFC superseded the role of the Honourable Irish Society with regard to fishery management and fishing rights within Lough Foyle which were purchased by the two governments under the auspices of the Commission for around £100,000 (Leary, 2016; NIA, 2010).

[8] Articles 2 and 3 of the Government of Ireland Act defined the partition based on constituency boundaries.

After decades of civil unrest on the IOI, the Good Friday Agreement[9] (GFA) in 1998 was lodged at the UN as an international agreement for peace among political parties in NI, as well as the Irish and British Governments. The advent of the GFA meant that ROI's constitutional claim to all the waters around the IOI was removed and in effect a notional international boundary was introduced between the two countries (Byrne, 1998). Articles 2 and 3 as noted above were amended to affirm that the Irish nation is a community of individuals with a common identity (not a territory) and unification (not re-integration), or a 'United Ireland' can only be 'brought about only by peaceful means with the consent of most of the people, democratically expressed, in both jurisdictions on the island' (Nineteenth Amendment of the Constitution Act 1998). Unlike Carlingford Lough, the UK Government claims ownership for the whole of Lough Foyle (up to the high-water mark on the Donegal side) (Seanad Eireann, 2006). In parallel, the Irish Government rejects the UK's claim regarding ownership of Lough Foyle (Seanad Eireann, 2006).

The GFA mitigated the lack of maritime boundaries by giving unique cross-border remits to fisheries, navigation and safety, marine tourism to various bodies. The creation of the Foyle, Carlingford, and Irish Lights Commission (FCILC) and Loughs Agency (LA)[10] in 1998 via the GFA replaced the FFC. The LA's remit includes fisheries management and marine tourism for river catchments and marine areas in Carlingford Lough and Lough Foyle.[11] The inability to resolve border issues within these two Loughs has legislative and policy implications for marine planning and management. Figure 3 illustrates the increased resource use conflicts and missed opportunities in the cross-border Loughs from the year 2000. One example is project Kelvin which had to be rerouted

[9] Originally known as the Belfast Agreement, signed between the British and Irish governments, political parties in NI to end that ended most of the political conflict on the governance of NI. It is known as the GFA because it was finalised on Good Friday, 10 April 1998.

[10] The Loughs Agency is sponsored by the Department of Agriculture, Environment and Rural Affairs (DAERA) in NI and the Department of the Environment, Climate and Communications (DECC) in the ROI has fisheries protection functions including the management of inland fisheries of the Foyle and Carlingford areas.

[11] The North/South Cooperation (Implementation Bodies) (NI) Order 1999.

and tension over the approval of CO_2 tank terminal at Warrenpoint in Carlingford.

The LA is prevented from exercising statutory powers over marine aquaculture licensing because of the contested ownership of the Lough. The 1952 Foyle Fisheries Act focused on licensing and regulation of salmon and trout fisheries on the Lough and made no specific reference to shellfish and aquaculture. The Foyle and Carlingford Fisheries (NI) Order 2007 and the Foyle and Carlingford Fisheries Act (ROI) 2007 provided a new aquaculture regulatory and proposed licensing system.

These two pieces of legislation covered the transfer of existing aquaculture licensing provisions from responsible departments to the FCILC and by default the LA. Yet, the failure to implement the full provisions of the legislations due to the ownership and jurisdictional issues has stalled efforts towards having a management agreement between the UK government, Irish Government, and the Crown Estate for the LA to have the remit to provide licences for shellfish (Ansong et al., 2022). This has led to unlicensed oyster trestles which have grown from around 2500 in 2010 to around 50,000 in 2018 with an estimated worth of £20 million if properly licensed (House of Commons, 2018).

A Memorandum of Understanding on lease of the seabed for offshore renewable energy installations was signed between both Governments in 2011, this came after the failed Tunes Plateau Offshore Wind Farm proposal (Northern Ireland Office, 2011). The Tunes Plateau Offshore Wind Farm was proposed between 2005 and 2010 on the north coast of NI but was withdrawn due to confusion over competencies (Ritchie et al., 2019). Other projects, such as the Kelvin submarine telecommunication cable, were also rerouted in 2009 to Coleraine (instead of Derry - Londonderry City) due to border disputes, unclear jurisdictional competencies, and political disputes between nationalist and unionist parties (NIA, 2009).

The next section discusses how these jurisdictional issues around borders and related spatial conflicts are considered in MSP for the IOI and Pomeranian Bay.

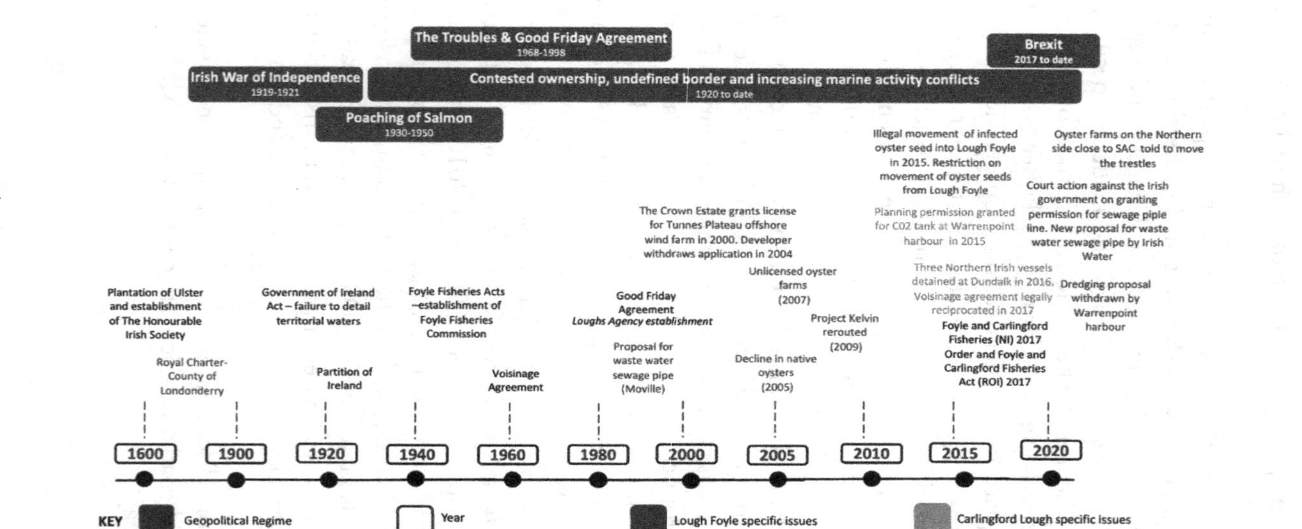

Fig. 3 Timeline of key historic conflicts between activities, planning, and management issues in Carlingford Lough and Lough Foyle. Produced by the author

Irresolution of Contested Borders and Spatial Conflicts in MSP

The case studies presented above highlight two different approaches to considering contested borders in MSP which are discussed below.

Opposing Representation of a Contested Border in the Pomeranian Bay

In the Pomeranian Bay, three different marine plans cover the contested area. These were prepared by five different leading authorities: the Federal Maritime and Hydrographic Agency for the German EEZ, the Ministry for Energy, Infrastructure and Digitalisation for the state of Mecklenburg-Vorpommern in Germany, and the Directors of Maritime Offices in Gdynia, Słupsk, Szczecin in Poland. In Germany, MSP is organised according to its federal structure with competences and functions distributed across State and Federal levels with each performing a distinct function. The second-generation MSP for the German EEZ in the North and Baltic Sea was adopted in 2021, whilst the Spatial Development Programme Mecklenburg-Vorpommern (LEP MV) which borders the Pomeranian Bay was adopted in 2016. The Polish MSP system follows a national plan adopted in 2021 covering the entire EEZ, territorial sea, and internal sea waters apart from ports and lagoons.

The German EEZ plans and the Polish MSP present similar visions including socio-economic development, environmental protection, and addressing spatial conflicts. The Polish MSP covers more sectors such as coastal protection, culture and heritage, reserve for future development, whilst the German EEZ covers less sectors and sub-sectors. The textual part of the Polish plan includes general provisions for each sector, detailed decisions about designation of specific basins and information on significant conditions influencing future use of each basin, and distribution of public investments. Similarly, the text of the German EEZ plan details out the objectives, principles, and justification (requirements and conditions for use) for each sector. Although both plans take a zoning approach to designate maritime uses, the Polish plan is more detailed as it divides the Polish maritime jurisdiction into 95 basins and 428 sub-basins (see Fig. 4). The plan area is assigned a unique letter code POM, and the basins were assigned subsequent unique numbers (e.g., POM.01.Ip in the case of the contested area).

In Germany, there is a slight difference in the approaches taken by the EEZ and Länder plans. The German EEZ spatial designation is according to priority and reservation areas where:

- Priority areas are reserved for a defined use. In this case, other uses conflicting with priority use are excluded;
- Reservation areas are reserved for uses given special weight in the planning process. A case-by-case decision is made here to determine which other uses may be permitted through a comparative analysis based on their significance and lack of acceptable alternatives.

In contrast, the LEP MV defines three spatial designation including priority, reservation areas, and a third called suitable areas. The suitable areas are reserved areas where specific uses are permitted and precluded elsewhere.

Both the German and Polish plans are coherent from a planning philosophy perspective as they apply a zoning and regulatory approach. Critically, there are differences in how the contested area (215.76 square kilometres) is represented. Whilst German plans use a fuzzy approach without designating uses for the contested area, the Polish plan makes a port or haven function designation to ports in Szczecin and Świnoujście for the contested area (POM.01.Ip). This led to different MSP approaches for the contested area and subsequent tension between planners in the two jurisdictions (Ansong et al., 2023). Overall, the differences in MSP approaches for the contested area were not resolved during the MSP process.

Fig. 4 The Maritime Spatial Plan of Polish Sea Areas. Source: Ansong et al. (2023: 2458), reproduced under the Creative Commons Attribution License [http://creativecommons.org/licenses/by/4.0/]

Discretionary and Fuzzy Approach to Borders on Island of Ireland

The National Marine Planning Framework (NMPF) was adopted in July 2021 as Ireland's first national framework for planning marine activities. The Department of Agriculture, Environment and Rural Affairs (DAERA), under the Marine Act (NI) 2013, has prepared a draft Marine Plan NI (dMPNI) in 2018 for inshore and offshore waters. The dMPNI is yet to be adopted as MSP currently sits at the bottom of the priority list for the NI Government amidst power struggles between unionist and nationalist parties, Brexit, and the COVID-19 pandemic. There are coherence points for both plans regarding the strategic, policy-based, and discretionary approach adopted by both jurisdictions. This is mainly based on cascading policies where projects must or should demonstrate that they will either avoid, minimise, or mitigate adverse impact on the environment. There are differences, the NMPF sets out and defines an overarching Transboundary Policy which emphasises the importance of transboundary consultation for projects that might have transboundary environmental impacts. Whilst the dNIMP does not define a specific transboundary policy, appropriate transboundary consultations are referenced under Air Pollution, Coastal Processes, Water Quality policies to be considered where projects may have possible transboundary effects. The NMPF allows for the development of regional and sectoral plans known as Designated Marine Area Plans (DMAPs), e.g., for offshore renewable energy.

The unregulated oyster farms in Lough Foyle are not presented in activity maps and not considered by either plan. Both plans identify the LA as a relevant body to facilitate cross-border engagement and implementation. The specific role of the LA in facilitating such roles in regarding a management agreement is not defined. Both plans do not define specific policy and map out sectoral conflicts and issues in Lough Foyle and Carlingford Lough. The spatial aspect of both plans is limited to existing uses and provides guidance and policies on how future proposals should be assessed during decision making.

The jurisdictional border that lies in the cross-border Loughs in both plans stops at the territorial sea between NI and the ROI (see Fig. 5).

Critically, both plans presented vague transboundary policies to address issues in both Loughs and jurisdictional issues were simply noted as outstanding and unresolved. For example, the NMPF states that:

> The resolution of jurisdictional issues in Lough Foyle and Carlingford Lough remains outstanding. Following discussions in 2011 between the Minister for Foreign Affairs and the British Foreign Secretary, the UK and Irish Governments agreed to seek to resolve these issues. Since then, a series of meetings have taken place at official level between the Department of Foreign Affairs and the Foreign and Commonwealth Office. The issues involved are complex and involve a range of different actors, including the Crown Estate, and both Governments are committed to achieving a positive resolution as soon as possible. (DHLGH, 2021: 22)

Overall, first-generation marine plans on the IOI merely sets high level Government's vision, objectives, and marine planning policies for marine activities. In the contested cross-border Loughs, the implementation of MSP is challenged by the lack of ambition by both governments to address ownership issues, tensions between stakeholders on both sides of the border, environmental issues, resource use conflicts, and limited legislative remits for the LA to administer license for marine aquaculture.

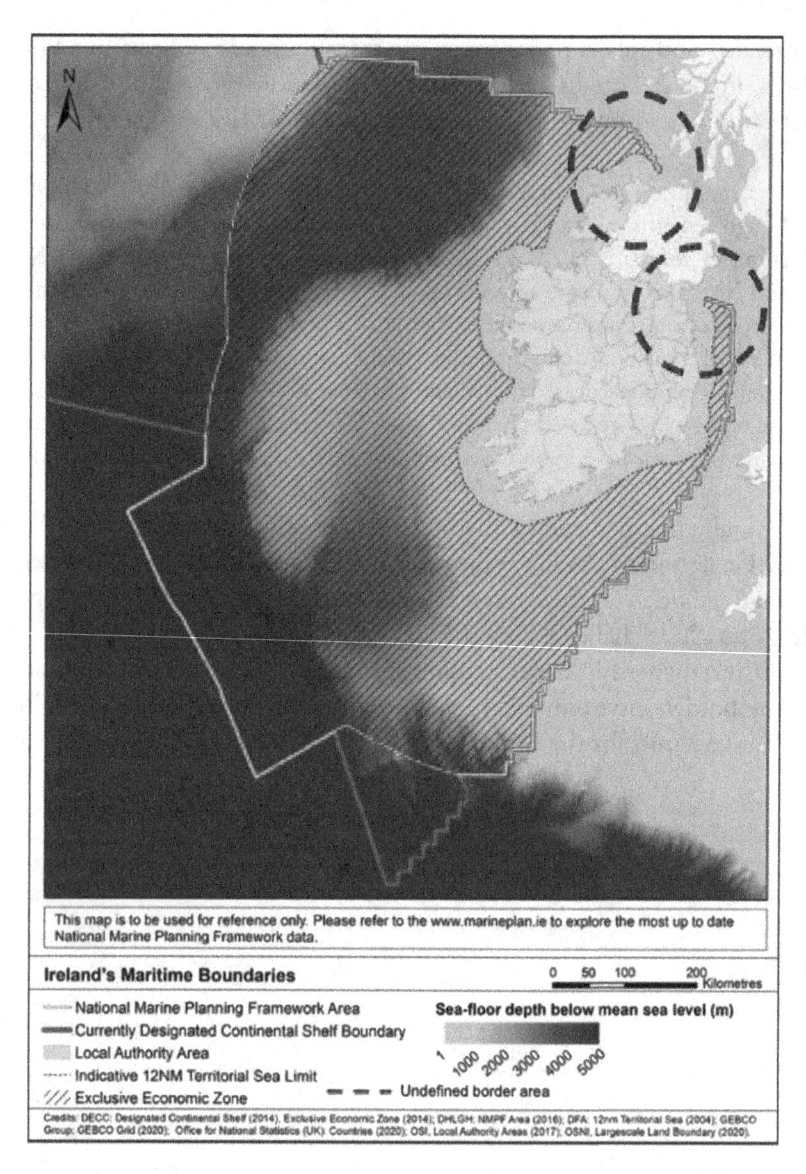

Fig. 5 Republic of Ireland Marine Plan Area. Source: Ansong et al. (2023: 2454), reproduced under the Creative Commons Attribution License [http://creativecommons.org/licenses/by/4.0/]

Conclusion

This chapter explores how practices of Marine Spatial Planning (MSP) factor in contested national borders. It further assessed the effectiveness of MSP in addressing spatial conflicts in two contested marine areas: Pomeranian Bay and the IOI. The two contested regions are significant, in that they reflect nations that have a matured MSP process which is expected to address contested resource conflicts.

The results from this review highlight two different approaches for considering contested borders in MSP: (i) the zoning/contradictory approach used for the Pomeranian Bay case, and (ii) the fuzzy boundary/ discretionary approach used on the IOI. For the Pomeranian Bay and especially the Polish plan, MSP became a process to make a political statement and introduce a long-held stance and legacy on the contested border. The Polish plan makes a port/haven function designation (POM.01.Ip) for the contested area to ensure access to the Szczecin and Świnoujście ports. In contrast, the German EEZ plan and LEP MV use a fuzzy approach (like the IOI example) without necessarily designating uses for the contested area. This led to different MSP approaches for the contested area and subsequent tension between planners. It could be argued that access to the Pomeranian Bay is of a higher national importance to Poland based on the location of the seaports for trade and commercial purposes. Hence, their drive is to ensure that such use of the maritime space is secured in a government approved MSP. Alternatively, environmental issues and tourism activities (on the German side) do not feature highly on the political agenda for the German Government, hence the fuzzy approach to addressing the border issue.

Without a proper agreement by both jurisdictions, there is the tendency for a 'territorial trap' to occur and potential exclusion from the use of space. For example, in the Pomeranian Bay, there were incidents of the prohibition of movement of vessels with the German navy stopping vessels going to Polish ports. MSP in this case became a process for continuing the 'territorial trap' approach and division between planners and users without extending flexible use of the area as noted by authors such as Peters (2020), Faludi (2019), and Nail (2016).

For the IOI case, MSP was a process that continued the political inaction on the contested border by introducing a discretionary approach. The discretionary nature of the marine plans ensures that the NI and ROI approach to MSP is less technical and more flexible, yet rarely identifies or addresses spatial conflicts in the contested area. Planning for spatial conflicts in the cross-border Loughs was noted as 'outstanding' (i.e., absent) by both plans. Although the IOI approach promotes a more flexible approach to the management of contested areas, the limited action and political backing have made such attempts difficult to implement. This has not led to uses of sea space being excluded, and hence the theory of territorial traps does not necessarily apply here. However, marine users have taken advantage of the discretionary nature and the deadlock between both countries to exploit marine resources without appropriate planning, e.g., unregulated development of oyster farms.

Overall, both approaches did not deliver effective planning for contested marine waters—which remains a 'planners dream' that is not achievable. As noted above, borders and jurisdictional issues are historically shaped, and overtime are engrained in how institutions work, impacting MSP. The results show that land borders are normally prioritised and agreed between countries over maritime borders. But where maritime borders apply, we are stuck with landed models of demarcation, meaning the flexibility the sea may need cannot always be realised in management practices. This further resonates with argument by Peters (2020) that so many of our approaches to marine management are terrestrial based. The sea, although valued by people, still remains removed from the environments of daily decision makers and beyond the consideration of more flexible approaches. Border or no border, MSP should apply an ecosystem-based approach which includes enhancing connectivity between marine ecosystems, users, and institutions. The introduction or absence of borders should not lead to exclusion of users or uncontrolled development. There is no need for countries to have the same planning systems or governance architecture. I argue in this chapter that there should be mechanisms and capacity in place to ensure that there is a balancing act and engagement to align processes to effectively manage contested marine waters.

The introduction of flexible approaches to MSP and marine management should have a strong focus on limiting social divisions, a clear legal basis, and provide capacity for national and cross-border institutions managing it to address contested issues. This highlights the role of bridging organisations and introducing a collaborative governance framework in cross-border areas for institutions and users to engage beyond borders to appreciate connectivity and collaborative working.

References

Acemoglu, D., Johnson, S., & Robinson, J. (2005). The rise of Europe: Atlantic trade, institutional change, and economic growth. *American Economic Review, 95*(3), 546–579.

Alexander, L. M. (1986). The delimitation of maritime boundaries. *Political Geography Quarterly, 5*, 19–24.

Ansong, J. O., Ritchie, H., Gee, K., McElduff, L., & Zaucha, J. (2023). Pathways towards integrated cross-border marine spatial planning (MSP): Insights from Germany, Poland, and the island of Ireland. *European Planning Studies, 31*(12), 2446–2469.

Ansong, J. O., McElduff, L., & Ritchie, H. (2021). Institutional integration in transboundary marine spatial planning: A theory-based evaluative framework for practice. *Ocean & Coastal Management, 202*, 105430. https://doi.org/10.1016/j.ocecoaman.2020.105430

Ansong, J. O., Ritchie, H., & McElduff, L. (2022). Institutional barriers to integrated marine spatial planning on the Island of Ireland. *Marine Policy, 141*, 105082. https://doi.org/10.1016/j.marpol.2022.105082

Boemeke, M., Feldman, G., & Glaser, E. (1998). *The Treaty of Versailles: A reassessment after 75 years (Publications of the German Historical Institute).* Cambridge University Press.

Byrne, D. (1998). An Irish view of the Northern Ireland peace agreement. The interaction of law and politics. *Fordham International Law Journal, 22*, 1206. https://ir.lawnet.fordham.edu/ilj/vol22/iss4/7

Department of Housing, Local Government and Heritage (DHLGH). (2021). *National marine planning framework*. Retrieved 29 May, 2024, from https://www.gov.ie/pdf/?file=https://assets.gov.ie/139100/f0984c45-5d63-4378-ab65-d7e8c3c34016.pdf#page=null

Fairbanks, L., Campbell, L., Boucquey, N., & St. Martin, K. (2018). Assembling enclosure: Reading marine spatial planning for alternatives. *Annals of the American Association of Geographers, 108*(1), 144–161.

Faludi, A. (2013). Territorial cohesion, territorialism, territoriality, and soft planning: A critical review. *Environment & Planning A: Economy and Space, 45*(6), 1302–1317.

Faludi, A. (2019). New horizons: Beyond territorialism. *Europa XXI, 36*, 35–44.

Flannery, W., O'Hagan, A., O'Mahony, C., Ritchie, H., & Twomey, S. (2015). Evaluating conditions for transboundary marine spatial planning: Challenges and opportunities on the Island of Ireland. *Marine Policy, 51*, 86–95.

Focus. (2013). The history of the curzon line: How did Stalin become a defender of the polish cause? Retrieved December 7, 2021, from https://www.focus.pl/artykul/historia-linii-curzona-jakstalin-zostal-obronca-sprawy-polskiej

Forbes, V. L. (2001). *Conflict and cooperation in managing maritime space in semi-enclosed seas.* University of Chicago Press Economics Books.

House of Commons. (2018) *Northern Ireland Affairs Committee. Oral Evidence: Brexit and Northern Ireland Fisheries.* HC 878: UK Parliament.

Jackowska, N. (2008). The border controversy between the Polish People's Republic and the German Democratic Republic in the Pomeranian Bay. *Przegląd Zachodni, 3*, 145–159.

Jay, S. (2018). The shifting sea: From soft space to lively space. *Journal of Environmental Policy and Planning, 20*(4), 450–467.

Kuempel, C. D., Jones, K. R., Watson, J. E., & Possingham, H. P. (2019). Quantifying biases in marine-protected-area placement relative to abatable threats. *Conservation Biology, 33*(6), 1350–1359.

Lambach, D. (2021). The functional territorialization of the high seas. *Marine Policy, 130*, 104579. https://doi.org/10.1016/j.marpol.2021.104579

Leary, P. (2016). *Unapproved routes: History of the Irish Border, 1922–1972.* Oxford University Press.

Maier, C. S. (2016). *Once within borders.* Harvard University Press.

Maxwell, S. M., Hazen, E. L., Lewison, R. L., Dunn, D. C., Bailey, H., Bograd, S. J., Briscoe, D. K., Fossette, S., Hobday, A. J., Bennett, M., Benson, S., Caldwell, M. R., Costa, D. P., Dewar, H., Eguchi, T., Hazen, L., Kohin, S., Sippel, T., & Crowder, L. B. (2015). Dynamic Ocean management: Defining and conceptualizing real-time management of the ocean. *Marine Policy, 58*, 42–50.

Nail, T. (2016). *Theory of the border*. Oxford University Press.
Northern Ireland Office. (2011). *Memorandum of understanding: Offshore renewable energy.* Memorandum of Understanding (MoU) between UK and Ireland Governments on offshore renewable energy development. Northern Ireland Office.
Northern Island Assembly (NIA). (2010). *Lough Foyle – Ownership, Licensing, and Levy Issues Relating to Shellfish.* Research and Library Service Briefing Paper. Retrieved December 7, 2021, from https://www.niassembly.gov.uk/globalassets/documents/raise/publications/2017-2022/2021/infrastructure/54212.pdf
Northern Island Assembly (NIA). (2009). *Official report (Hansard), Project Kelvin.* Retrieved December 7, 2021, from http://www.niassembly.gov.uk/assembly-business/official-report/committeeminutes-of-evidence/session-2008-2009/february-2009/project-kelvin/
O'Dowd, L. (2010). From a 'borderless world' to a 'world of borders': 'Bringing history back in.' *Environment and Planning D: Society and Space, 28*(6), 1031–1050.
Office of the Attorney General. (1937). *Irish Constitution 1937.* Retrieved December 7, 2021, from https://www.irishstatutebook.ie/eli/cons/en/html
Østhagen, A. (2019). *Lines at sea: Why do states resolve their maritime boundary disputes?* (Doctoral Dissertation). University of British Columbia.
Peters, K. (2020). The territories of governance: Unpacking the ontologies and geophilosophies of fixed to flexible ocean management, and beyond. *Philosophical Transactions of the Royal Society B, 375*, 20190458. https://doi.org/10.1098/rstb.2019.0458
Ritchie, H., Flannery, W., O'Hagan, A. M., Twomey, S., & O'Mahony, C. (2019). Marine spatial planning, Brexit, and the island of Ireland. *Irish Geography, 52*(2), 213–233.
Seanad Eireann. (2006). *Irish Government: Parliamentary debates. volume 248, no. 13.* Retrieved June 6, 2024, from https://www.oireachtas.ie/en/debates/debate/seanad/2006-06-13/
Ślepowroński, T. (2005). GDR Vs PRL, Relation of West Pomeranian inhabitants to the conflict in The Pomeranian Bay (1985–1989). *Biuletyn Instytutu Pamięci, 5*(9/10 (56/57)), 90–99.
Steinberg, P. E. (1999). Lines of division, lines of connection: Stewardship in the world ocean. *Geographical Review, 89*(2), 254–264.

Symmons, C. (2009). The maritime border areas of Ireland, north and south: An assessment of present jurisdictional ambiguities and international precedents relating to delimitation of border bays. *International Journal of Marine and Coastal Law, 24*(4), 57–500.

Twomey, S. (2020). *Seeking pathways towards improved transboundary environmental governance in contested marine ecosystems* (Doctoral Dissertation). University College Cork.

United Nations (UN). (1951). *Demarcation of the State Frontier between Germany and Poland signed in Frankfurt Oder on 27 January 1951.* United Nations.

United Nations (UN). (1989). *Treaty between the German Democratic Republic and the Polish People's Republic on the Delimitation of the Sea Areas in the Oder Bay 22 May 1989.* United Nations. Retrieved December 7, 2021, from https://www.un.org/depts/los/LEGISLATIONANDTREATIES/PDFFILES/TREATIES/DEU-POL1989DS.PDF

United Nations (UN). (1990). *Treaty between the Federal Republic of Germany and the Republic of Poland on the confirmation of the frontier between them, 14 November 1990.* United Nations. Retrieved December 7, 2021, from https://www.un.org/depts/los/LEGISLATIONANDTREATIES/PDFFILES/TREATIES/DEU-POL1990CF.PDF

United Nations (UN). (2021). *The second world ocean assessment II*. United Nations.

Imaginaries: Oceanic Bordering with Large-Scale Marine Protected Areas

Jasper Montana and Oscar Hartman Davies

Introduction

In 2016 at the World Conservation Congress of the International Union for the Conservation of Nature (IUCN) in Hawaii, the President of the Republic of Palau announced plans for a marine protected area that would cover 80% of the Pacific Island nation's oceanic waters. In the address, the President H.E. Tommy E. Remengesau, Jr. reflected that:

J. Montana (✉)
Centre for the Public Awareness of Science, Australian National University, Canberra, Australia
e-mail: jasper.montana@anu.edu.au

J. Montana · O. H. Davies
School of Geography and the Environment, University of Oxford, Oxford, England
e-mail: oscarhd@kth.se

O. H. Davies
KTH Royal Institute of Technology, Stockholm, Sweden

K. Peters and J. Turner (eds.), *Ocean Governance (Beyond) Borders*,
https://doi.org/10.1007/978-3-031-71322-4_6

> in Palau, stewardship of the land and the oceans are the touchstones that have enabled our people to thrive for millennia. Ladies and Gentlemen, the Republic of Palau, as a Large Ocean State, takes the responsibility of conservation very seriously and … [I have] had the privilege of signing into law, one of the largest marine protected areas in the world. … The Palau National Marine Sanctuary is our contribution to the global effort to restore our oceans. It is also our commitment to our children's future. (Remengesau Jr., 2016, September 1)

The Palau National Marine Sanctuary came into effect in January 2020. It is one example of a growing number of large-scale marine protected areas (LSMPAs)—typically defined as those covering areas greater than 150,000 square kilometres. LSMPAs are increasingly being established worldwide and now make up a substantial proportion of marine space under protection (Big Ocean, 2017).

The establishment of LSMPAs is a particular moment in which bordering regimes and the imagination intersect with regard to ocean space. While LSMPAs are considered most for their potential for conserving biodiversity or enabling sustainable fisheries, the Palauan President's statement about this new marine sanctuary illustrates that protected areas in the ocean involve more than technical concerns. Rather, as this chapter will show, the bordering practices of protected areas are also bound up in collectively-held ideals about desired futures—and remembered pasts—for the nation states that enact them. In this chapter, we outline how this understanding expands the recognised role of imagination in ocean governance and complements thinking about ocean governance beyond borders.

The chapter begins with a discussion of the recent rise of LSMPAs as an increasingly common approach to ocean governance and positions this trend in an emerging literature on imagination and the ocean. The chapter reflects on two cases of LSMPA designation: first, in the Republic of Palau; and second, in the overseas territories of the UK. Through an analysis of political discourse surrounding their designation, we explore the way that the creation of LSMPAs is also tied up with imaginations of the identity of these nation states, their people and their place in the world. Recognising the contribution of imagination to ocean governance that extends beyond the material and spatial imagination of the ocean itself is crucial to understanding how and why the bordering logics of LSMPAs are so attractive for the states that enact them. The bordering of ocean space by nations with large marine estates enables them to reinscribe and reinvent cultural ties to the ocean while also serving broader international conservation goals.

The Rise of LSMPAs

With many new marine protected areas of more than 150,000 square kilometres designated or inscribed since 2010, the creation of LSMPAs is on the rise (see Table 1). The first LSMPA to be recognised was the Great Barrier Reef Marine National Park (346,000 square kilometres) established in 1975 in Australia. Since then, there has been a proliferation of LSMPA designations largely amongst the remote overseas territories of former colonial powers and across independent Pacific Island states. This ocean governance strategy typically rests on the drawing of ocean lines—or borders—extending 200 nautical miles from land to the edges of a country's Exclusive Economic Zones (EEZ). The authority of governments to designate LSMPAs derives from the UN Convention on the Law of the Sea (1982), which assigns their rights to exclusively explore, exploit and manage undersea resources within their EEZs without conveying sovereign territorial ownership of these otherwise international waters. The potential to enact such enclosures beyond the EEZs of individual nation states was further enhanced in March 2023 with agreement of the 'High Seas Treaty': a United Nations agreement that for the first time establishes a framework for governments to agree on area-based protection measures in the high seas (Stokstad, 2023).

LSMPAs underpin a global agenda for area-based marine protection oriented towards addressing marine biodiversity loss and the 'better' management of global fisheries (Campbell et al., 2021). Since 2001, governments have used the international Convention on Biological Diversity as a mechanism to set a global target of designating 20% of land and 10% of the sea as protected. In December 2022, the Kunming-Montreal Global Biodiversity Framework further ramped up

Table 1 Large-scale marine protected areas (greater than 150,000 km^2) designated or inscribed within Exclusive Economic Zones worldwide between 2010 and 2020

Name	Reported area[a]	Status	Status year	Responsible government
Tristan da Cunha	687,223	Designated	2020	UK
Amirantes (Marine) to Fortune Bank (Marine) Area of Outstanding Natural Beauty	217,589	Designated	2020	Seychelles
Aldabra Group (Marine) National Park	201,236	Designated	2020	Seychelles
French Austral Lands and Seas	672,969	Inscribed	2019	France
Ascension Island Marine Protected Area	445,000	Designated	2019	UK
Tuvaijuittuq Marine Protected Area	319,411	Designated	2019	Canada
Coral Sea	684,956	Designated	2018	Australia
Rapa Nui	579,368	Designated	2018	Chile
Área De Proteção Ambiental Do Arquipélago De Trindade E Martim Vaz	403,854	Designated	2018	Brazil
Área De Proteção Ambiental Do Arquipélago De São Pedro E São Paulo	384,133	Designated	2018	Brazil
Mar de Juan Fernández	264,443	Designated	2018	Chile
Coral Sea	238,400	Designated	2018	Australia
Marae Moana (Cook Islands Marine Park)	1,968,938	Designated	2017	Cook Islands
Papahanaumokuakea Marine	1,508,858	Designated	2016	USA
Pitcairn Islands	841,910	Designated	2016	UK
Pacífico Mexicano Profundo	436,141	Designated	2016	Mexico
Nazca-Desventuradas	300,035	Designated	2016	Chile
Palau National Marine Sanctuary	502,538	Designated	2015	Palau

(continued)

Table 1 (continued)

Name	Reported area[a]	Status	Status year	Responsible government
Natural Park of the Coral Sea	1,292,967	Designated	2014	France
Prince Edward Island Marine Protected Area	181,229	Designated	2013	South Africa
South Georgia and South Sandwich Islands Marine Protected Area	1,240,000	Designated	2012	UK
British Indian Ocean Territory Marine Protected Area (Chagos)	640,000	Designated	2010	UK
Phoenix Islands Protected Area	408,250	Inscribed	2010	Kiribati
Papahānaumokuākea	362,075	Inscribed	2010	USA
Motu Motiro Hiva	150,000	Designated	2010	Chile

[a] Rounded to nearest square kilometre. Data source: UNEP-WCMC & IUCN, 2024

this ambition by agreeing to protect at least 30% of both land and sea areas by 2030 (CBD, 2022). LSMPAs appear as an essential strategy for the attainment of these targets over and in addition to the piecemeal protection of smaller marine areas as part of a new wave of 'ocean enclosure' (Boucquey et al., 2019; Campbell et al., 2016). Although LSMPAs represent a small number of more than 18,000 marine protected areas globally, they contribute significantly to the area of ocean under protection (UNEP-WCMC & IUCN, 2024). Indeed, the growth in LSMPA designation reflects a 'bigger-is-better' mentality that has been strongly promoted by major environmental NGOs, including The Pew Charitable Trusts, The Nature Conservancy, and the National Geographic Society (Campbell et al., 2016; Smyth & Hanich, 2019).

Beyond achieving global goals on biodiversity and sustainable development, however, the creation of LSMPAs is clearly also a geopolitical concern (Campbell et al., 2021; De Santo, 2020; Gray, 2018). At their heart, LSMPAs are territory-building technologies that expand a nation state's active sovereignty over ocean space through bordering, monitoring and enforcing human activities. The authority to establish LSMPAs has existed for decades but the rapid increase in new LSMPAs has been partly enabled by the improved affordability and advancement of satellite imagery and other remote sensing technologies, which have increased capacity to carry out monitoring and enforcement of these large areas of ocean (Bakker, 2022). The protection afforded from fisheries and other extractive industries by recently established LSMPAs varies however. The framework of the International Union for the Conservation of Nature (IUCN) supports a range of different management options for marine protected areas, from completely 'no take' (IUCN Category I) to 'sustainable use' where economically important fishing is allowed (IUCN Category VI). Many LSMPAs also include inshore regions that allow for the continued use of marine resources for subsistence purposes by local communities. For instance, the Marae Moana Marine Park covers the entire EEZ of the Cook Islands, but as a multiple-use marine park commercial fishing is allowed outside of no-take areas that cover the 50 nautical mile radius of each of the state's islands (Cook Islands Government, 2023).

Yet, the economic opportunities of marine resource exploitation coupled with the costs of monitoring and enforcement in marine protected areas can create challenges in some cases (Farran, 2022). For example, in Kiribati, economic revenue from licenced tuna fishing resulted in plans to return commercial fisheries to the Phoenix Islands Protected Area in 2021 (Farran, 2022), and consultation has also taken place concerning the licensing of deep-sea mining operations for a rich deposit of manganese nodules in the Marae Moana Marine Park of the Cook Islands (Farran, 2022). These economic and environmental tensions raise important questions concerning the longevity, differential capacities and costs of maintaining LSMPAs. Despite these concerns, LSMPAs have become an important feature of the global governance regime for ocean space in the twenty-first century (Campbell et al., 2016).

Imagination and the Ocean

Imagination plays a key role in contemporary environmental governance (Chhetri et al., 2022), and has long been mobilised for the control of both people and place through the deployment of knowledge and power for political purposes (Braun, 2015; Gregory, 1995; Said, 1993). In this chapter, we understand imagination in environmental governance as it pertains to specific visions that are evoked in the public arena to advocate for change and legitimate policy choices through the mobilisation of discourses, metaphors, visual images, maps and so on. The role of imagination in the governance of the ocean is particularly significant because, in many present-day societies, the governing elite—and indeed many of their inhabitants—have limited physical contact with the ocean, making it a place mostly accessed through imagination (see, e.g., Campbell et al., 2021; Crawley et al., 2022; Steinberg, 2008).

How, then, is the ocean imagined? The answer to this question—of course—changes in different times and places, and amongst different cultures and communities. Historically, for example, North Atlantic merchants, seafarers and governments have imagined ocean space as a functional space for international trade offering 'connection and an arena of mobility' (Steinberg, 2009: 468), or as an unruly 'outside' that sits in contrast to the orderly 'universe of state-civilizations' (ibid: 472). Today, academic knowledge frameworks are increasingly central to imagining ocean space, be it the economist's vision of fish stocks and quotas, or the anthropologist's vision of a peopled sea (Cardwell & Thornton, 2015). Recent scholarship has shown the dominance of an ocean governance logic centred on land-based imaginaries of ocean space. Geographers have noted how the use of lines and fences that are typically used on land to control access and police behaviour in terrestrial areas 'has become a blueprint approach for ocean governance' (Peters, 2020: 4). Steinberg and Peters (2015) describe this as a territorial ontology, in which ocean space is imagined as a horizontal extension of land that is able to be demarcated and partitioned for effective governance through borders. Such bordering practices are clearly operating in the creation of LSMPAs today. Indeed, this scholarship has shown that the ways in which ocean space is imagined has direct implications for the kinds of policies that are put in place to govern it (Peters, 2020).

To challenge the prominence of landed logics in ocean governance, scholars and others have sought to emphasise other ways of imagining ocean space that are less dependent on borders. Such approaches emphasise the way that terrestrial imaginaries are confounded by the materiality of the sea itself, where 'coastlines change, fish swim, water moves and ships travel' (Bear & Eden 2008, 487), requiring new watery imaginaries and, perhaps, more fluid, dynamic spatial understandings in marine governance (see Bakker, 2022; Bear, 2017; Peters, 2020). This 'wet ontology' is more appropriately attuned to the ocean as 'a world of flows, connections, liquidities, and becomings, [which includes] the reimagining and reenlivening of a world ever on the move' (Steinberg & Peters, 2015: 248). This perspective allows us to not only see the ocean as materially different to land as an imagined space for management, but also attune ourselves to recognise fluidity, volume and turbulence throughout spatial thinking (Peters, 2020). Connecting these geographical imaginations of ocean materiality with the perspectives of practitioners working close to the sea, Cardwell and Thornton (2015) have further called for recognition of the ocean as 'relational, heterogeneous and place based' (what they term a 'fisherly imagination'). Non-territorial geophilosophies are seemingly increasingly relevant to the practices of ocean governance. As new technologies afford greater capacities to monitor ocean space for the purpose of large-scale bordering projects, such as LSMPAs, they also paradoxically reveal the fluidity and dynamism of ocean life, including the mobilities and lifeworlds of marine species, fishing vessels and communities living close to the sea.

In this chapter, we build on these geographic perspectives in centring imagination as it plays out between societies and their relations to the ocean. To do so, we draw on conceptual resources from science and technology studies, where imagination has also gained increased attention. We work with the concept of sociotechnical imaginaries (Jasanoff, 2015), which foregrounds the importance of the imagination in the ordering of modern societies. This concept draws attention to 'visions of desirable futures, animated by shared understandings of forms of social life and social order attainable through, and supportive of, advances in science and technology' (Jasanoff, 2015: 4). This concept explicitly positions imagination as a tool of governance often 'grounded in positive visions of social progress' (Jasanoff, 2015: 4). As we outline in this chapter, what different nation states consider to be the 'right' relations with the ocean are intimately tied to their understanding of appropriate national visions of social progress. Bordering techniques, such as the creation of LSMPAs, are created to reinforce these imaginaries. We thereby move in the following section to articulate—through two specific examples—how imagination and the production of oceanic borders are intertwined in the designation of these LSMPAs.

The Cases of Palau and Ascension Island

In this section, we delve deeper into the relationship between imagination and ocean borders by focusing on two recently designated LSMPAs. The first is the Palau National Marine Sanctuary introduced at the start of this chapter. The Republic of Palau is a Pacific Island nation composed of over 300 islands, and its LSMPA was designated to cover 80% of the country's EEZ. The second example that we focus on is the UK and its Ascension Island Marine Protected Area established as part of the Blue Belt Programme, which sought to formalise protection of over four million square kilometres of ocean by 2020 across the UK Overseas Territories. Ascension Island is part of an overseas territory of the UK (currently classified as non-self-governing under Chapter XI of the Charter of the United Nations) situated in the Atlantic Ocean. The Ascension Island Marine Protected Area—designated in 2019 following an election manifesto commitment of the UK Conservative Party—covers 100% of its EEZ. These two cases illustrate the dominant trend of LSMPA creation, which are typically established by small independent island states or by former colonial powers. To explore these cases, we focus on imagination and the ocean as it is portrayed in the political discourse that arises at the time of LSMPA designation. While visual images and maps of the oceans are clearly significant resources for mobilising visions of ocean space (as considered in Campbell et al., 2021; Crawley et al., 2022), we do not focus on these in our account. Instead, we specifically draw on empirical material from political speeches and press-releases issued at the time of LSMPA designation, which emphasise the narratives of deep cultural ties to the ocean in the national identities of Palau and the UK. In this sense, our analysis is one of elite narratives that take us far from the fisherly imagination outlined by Cardwell and Thornton (2015). Yet, we contend that the worldmaking of political leaders needs to be also scrutinised to understand how global conservation agendas naturalise visions of ocean space. While both cases rely upon borders as a spatial logic in ocean governance, they provide contrasting visions of two nations that both see themselves as culturally connected to the sea.

Both Palau and the UK cases illustrate the way that imagination and new ocean borders come together within LSMPAs through the mobilisation of material, symbolic and cultural resources through which visions of social order are made durable (as per Jasanoff, 2015: 25). Although conservation goals remain firmly in the rhetoric surrounding the designation of LSMPAs, there are also a wider set of narrative at play that we explore here. The speech given by the President of Palau cited in the introduction to this chapter, for example, foregrounds themes related to national (and regional) identity from a distinct Pacific Island vantage point. Remengesau, Jr. (2016, September 1) emphasises that the establishment of an LSMPA in Palau therefore seeks to extend a long and rich history of convivial relations with the ocean as place and home, as well as a vision of Pacific Island nations as part of a family. His speech notes: 'Just as Palauans have always understood the importance of oceans, so too have our brothers and sisters across the Pacific. Together, we have been global conservation leaders, and I am proud of what we have accomplished together'. This aligns with the geophilosophies of Fijian and Tongan writer Epeli Hau'ofa (1994: 152), who has described 'a sea of islands': a large, fluid, and relational space in which the open ocean is a home. Home, Hau'ofa (1994: 159) continues, is accompanied by responsibilities of care and guardianship: '[t]here are no people on earth more suited to be the guardians of the world's largest ocean than those for whom it has been home for generations'. The designation of the Palau LSMPA is therefore arguably tied to an imaginary of social order that draws upon a rich cultural history connecting Pacific Islanders with the ocean and with each other. For this reason, the bordering practices of LSMPA designation in Palau are more than just producing new borders, but also erasing them.

The speech draws attention to the 'stewardship of the land and the oceans' by the people of Palau and their 'brothers and sisters' across the Pacific, challenging understandings of the ocean as 'outside', and kinds of national exceptionalism more evident in the UK example to come. Instead, the speech articulates the establishment of the Palau LSMPA alongside a commitment to nationhood with and through the sea, and also a regional vision of working politically and culturally with other Pacific Island nations to achieve shared aims. Similar narratives about Pacific unity have long-served political purposes in international forums, such as the United Nations. However, they are also recognised by Pacific studies scholars to belie differential cultural and political histories, as well as continuing tensions between the worldviews of mobile elites compared with diverse local communities (critically considered in Lawson, 2010; Ratuva, 2004). Relatedly, the speech by Remengesau, Jr. further asserts the identity of Palau as a 'Large Ocean State'. This is a term that is shown in previous scholarship to be mobilised by island states to emphasise their ocean-linked identity, and also enhance their negotiating power in international relations (Chan, 2018). In this way, the speech also hints towards the potential for LSMPAs to offer island states expanded forms of sovereignty underpinned by territorial logics of the ocean as planar space—a development that is particularly pertinent at a time when coastlines are experiencing inundation by rising seas due to climate change.

For the UK, there have been similar themes of national identity construction in the imaginary projected around LSMPA designation but tied to very different cultural histories of connection to the ocean. The designation of LSMPAs in the UK Overseas Territories has been drawn into a vision of British progress that aligns with a longer-term imaginary of Britain as a 'master' of the seas. In a speech to the 2016 *Our Ocean Conference*, for example, the UK Foreign Office Minister Alan Duncan explained: 'the value and significance we attach to marine protection is expressed in the scale of our ambition for the sea. ... Now we've become pioneers in marine protection just as we were, centuries ago, pioneers of discovery' (UK Foreign & Commonwealth Office, 2016, September 15). Similar narratives can be found in related UK Government press-releases, such as one entitled 'Britannia Protects the Waves' (UK Foreign & Commonwealth Office, 2019, August 26)—adapting text from the eighteenth-century poem 'Rule, Britannia' by James Thomson—and another that quotes then Foreign Office Minister Boris Johnson as stating: 'Britain has a proud and ancient maritime history, but our commitment to the oceans must be enshrined in our future' (UK Foreign & Commonwealth Office, 2018, June 22). Of course, this celebration of oceanic power as a so-called pioneer of discovery glosses over challenging aspects of the UK's complex and important seafaring history, including colonial rule and its roles in the atrocities of transatlantic slavery. However, these speeches suggest that the creation of LSMPAs is being seen by these political leaders as a means to project a new vision of social progress, in which the UK is a maritime power through its role as an environmental leader as part of a post-Brexit agenda of 'Global Britain' (Harmer et al., 2022). Critical scholars have contended, meanwhile, that the designation of ocean borders around the UK Overseas Territories reflects a colonial and militarised present. LSMPAs, they contend, can be seen as a means to reinforce the UK's territorial claims to sovereignty over the islands and their waters in the name of conservation (De Santo, 2020; De Santo et al., 2011).

The capacity to border ocean space through drawing lines on a map and monitoring these with 'Great British' technology at once affirms dominion over ocean space and projects a vision of UK scientific and technological progress. Indeed, the UK's Foreign Office Minister's speech goes on to assert that as part of the Blue Belt Programme, the UK would: 'like to see others join the UK's Ocean Innovation Hubs – a sort of set of Silicon Valleys for the oceans – to put this vision into practice' (UK Foreign & Commonwealth Office, 2016, September 15). Here, recent advances in the UK's scientific collaboration and technological innovation, including satellite systems for monitoring illegal fishing activities, are presented as central to advancing a particular desired future.

Both the Republic of Palau and the UK are vying to position themselves as world leaders in environmental stewardship through LSMPA designation. In doing so, they also seek to remember cultural connections to the ocean, albeit through arguably incomplete and inaccurate memories that leave difficult parts of these histories and disputed identities underexplored and underexplained. However, in both cases, these narratives ensure that the bordering of ocean space through LSMPAs presents itself as the logical action to take in the service of national progress, including an expanded interest in the economic potential of ocean space and its management as part of a 'blue economy'.

In both examples analysed here, cultural resources of past attachments to the ocean are evoked in putting forward visions of the future made possible through LSMPA designation. Yet, there are also clear inequalities between states, such as Palau and the UK, that should also be taken into account. While the expansion of territory through the bordering logics of LSMPAs is ostensibly available to any coastal state, the abilities to engage in ocean governance in this way are not universal and are limited by a state's access to appropriate resources. This aligns with a recognition that material resources also matter to the enactment of imagination as visions of progress are 'bound up with the hard stuff of past achievements' (Jasanoff, 2015: 22). The technologies and technical capacities needed to monitor and enforce the effectiveness of conservation interventions like LSMPAs, and the integrity of their borders, are not equally available to all (Bakker & Ritts, 2018). For nations such as the Republic of Palau clear tensions persist between the political priorities and economic costs of marine protection. Recent political discussions have seriously considered the rollback of aspects of protection for the LSMPA to generate further fishing-related revenue (Gunia, 2022). This illustrates how LSMPAs incur significant real and, in many cases, opportunity costs to the states that establish them, and for certain states these may outweigh capacity and political will. Implicit in the enactment of different imaginaries is the risk of a widening technological and economic divide whereby certain nation states are better resourced to create, monitor and enforce protected areas. This matters because LSMPAs are bold statements in ocean governance. As large-scale bordering achievements they put conservation action on the map by drawing ocean boundaries that can be seen from space. As such, compared with other conservation strategies, such as the temporal closure of fisheries in particular areas, they also have enhanced potential for international impact as part of a 'race to the top' in scaling up ocean protection to 30% by 2030 (Campbell et al., 2021). LSMPAs therefore not only enhance active sovereignty, but they can also shift the perceived geopolitical order by foregrounding the environmental stewardship capacities of the states that enact them. For this reason, continuing to understand how and why LSMPAs are being created and the politics that they produce and obscure remain important empirical questions worthy of further exploration.

Conclusion

In developing the concept of sociotechnical imaginaries, Jasanoff (2015: 17) describes imagination as 'a crucial reservoir of power and action, [which] lodges in the hearts and minds of human agents and institutions'. We contend that this is certainly the case for visions of progress in ocean governance. In this chapter, we have shown how LSMPAs serve not only a conservation and marine management purpose, but are mobilised as resources that inscribe visions of well-managed ocean space alongside visions of national identities tied to the ocean. Moreover, we have sought to articulate, through two specific examples, how scholars interested in these ocean-bordering practices (and their deficiencies) might benefit from deeper engagements with imagination as a tool, and object, of governance. In particular, there is value in examining how political discourses are mobilised to bring visions of the ocean into the public arena and to justify and facilitate the bordering of marine space in new ways. To this end, we broadly affirm recent calls to explore wet ontologies and fluid geophilosophies for ocean governance (Peters, 2020), as well as consider broader scholarship on imaginaries to unpack how ocean governance is entangled with past and present processes of state identity formation, colonial relations of dispossession and social control, and the significance of ocean connection to the ways of life of people in place.

Despite the plurality of imaginaries at play in environmental governance (Chhetri et al., 2022), attending to who has the resources to institutionally stabilise these visions of desirable futures is also important. As Chhetri et al., (2022: 3–4) remind us: '[i]magination is political

… and whose visions, aspirations, expectations, and values are incorporated within environmental governance is critical'. As such, scholarship and policy development will only benefit from recognising and—where desirable—enabling a wider range of imaginaries from different actors and across different scales in the development of future regimes for ocean governance. At issue here is not only the notion that LSMPAs, and other bordering practices undertaken in the name of marine protection, are about more than biodiversity conservation—although, certainly, this is the case. More significantly, the bordering practices that they entail are bound up with affirming and materialising particular social orders at the expense of others. As part of recognising the human dimensions of LSMPAs (called for in Gray et al., 2017; Gruby et al., 2016), such effects are worthy of greater attention.

References

Bakker, K. (2022). Smart oceans: Artificial intelligence and marine protected area governance. *Earth System Governance, 13*, 100141. https://doi.org/10.1016/j.esg.2022.100141

Bakker, K., & Ritts, M. (2018). Smart earth: A meta-review and implications for environmental governance. *Global Environmental Change, 52*, 201–211.

Bear, C. (2017). Assembling ocean life: More-than-human entanglements in the Blue Economy. *Dialogues in Human Geography, 7*(1), 27–31.

Bear, C., & Eden, S. (2008). Making space for fish: the regional, network and fluid spaces of fisheries certification. *Social & Cultural Geography, 9*(5), 487–504.

Big Ocean. (2017). *Large-scale marine protected areas: Guidelines for design and management*. IUCN. Retrieved September 1, 2023, from https://portals.iucn.org/library/sites/library/files/documents/PAG-026.pdf

Boucquey, N., Martin, K. S., Fairbanks, L., Campbell, L. M., & Wise, S. (2019). Ocean data portals: Performing a new infrastructure for ocean governance. *Environment and Planning D: Society and Space, 37*(3), 484–503.

Braun, B. (2015). Futures: Imagining socioecological transformation—An introduction. *Annals of the Association of American Geographers, 105*(2), 239–243.

Campbell, L. M., Gray, N. J., Fairbanks, L., Silver, J. J., Gruby, R. L., Dubik, B. A., & Basurto, X. (2016). Global oceans governance: New and emerging issues. *Annual Review of Environment and Resources, 41*(1), 517–543.

Campbell, L. M., Gray, N. J., Zigler, S. B. J., Acton, L., & Gruby, R. (2021). World-making through mapping: Large-scale marine protected areas and the transformation of global oceans. In M. Himley, E. Havice, & G. Valdivia (Eds.), *The Routledge handbook of critical resource geography* (pp. 425–440). Routledge.

Cardwell, E., & Thornton, T. F. (2015). The fisherly imagination: The promise of geographical approaches to marine management. *Geoforum, 64*, 157–167.

CBD. (2022). *Kunming-Montreal Global Biodiversity Framework (GBF)*. Convention on Biological Diversity. Retrieved September 1, 2023, from https://www.cbd.int/gbf/

Chan, N. (2018). "Large Ocean States": Sovereignty, small islands, and marine protected areas in global oceans governance. *Global Governance: A Review of Multilateralism and International Organizations, 24*(4), 537–555.

Chhetri, N., Ghimire, R., & Eisenhauer, D. C. (2022). Geographies of imaginaries and environmental governance. *The Professional Geographer, 75*(2), 263–268.

Cook Islands Government. (2023). *EEZ and fishing*. Cook Islands Government. Retrieved September 1, 2023, from https://www.maraemoana.gov.ck/about-marae-moana/eez-and-fishing/

Crawley, G., Critchley, E., & Neudecker, M. (2022). Imaginaries: Art, film, and the scenography of oceanic worlds. In K. Peters, J. Anderson, A. Davies, & P. E. Steinberg (Eds.), *The Routledge handbook of ocean space* (pp. 277–297). Routledge.

De Santo, E. M. (2020). Militarized marine protected areas in overseas territories: Conserving biodiversity, geopolitical positioning, and securing resources in the 21st century. *Ocean & Coastal Management, 184*, 105006. https://doi.org/10.1016/j.ocecoaman.2019.105006

De Santo, E. M., Jones, P. J. S., & Miller, A. M. M. (2011). Fortress conservation at sea: A commentary on the Chagos marine protected area. *Marine Policy, 35*(2), 258–260.

Farran, S. (2022). Deep-sea mining and the potential environmental cost of 'going green' in the Pacific. *Environmental Law Review, 24*(3), 173–190.

Gray, N. J. (2018). Charted waters? Tracking the production of conservation territories on the high seas. *International Social Science Journal, 68*(229–230), 257–272.

Gray, N. J., Bennett, N. J., Day, J. C., Gruby, R. L., Wilhelm, T. A., & Christie, P. (2017). Human dimensions of large-scale marine protected areas: Advancing research and practice. *Coastal Management, 45*(6), 407–415.

Gregory, D. (1995). Imaginative geographies. *Progress in Human Geography, 19*(4), 447–485.

Gruby, R. L., Gray, N. J., Campbell, L. M., & Acton, L. (2016). Toward a social science research agenda for large marine protected areas. *Conservation Letters, 9*(3), 153–163.

Gunia, A. (2022). One island nation's controversial plan to take climate justice into its own hands. *Time*. Retrieved September 1, 2023, from https://time.com/6171610/palau-marine-protected-area-climate-justice/

Harmer, N., Bailey, I., & Hart, N. (2022). UK state identity-making and British overseas territories' environments in times of ecological crisis and geopolitical change. *Small States & Territories, 5*(1), 31–54.

Hau'ofa, E. (1994). Our sea of islands. *The Contemporary Pacific, 6*(1), 148–161.

Jasanoff, S. (2015). Future imperfect: Science, technology, and the imaginations of Modernity. In S. Jasanoff & S.-H. Kim (Eds.), *Dreamscapes of modernity: Sociotechnical imaginaries and the fabrication of power* (pp. 1–33). University of Chicago Press.

Lawson, S. (2010). 'The Pacific Way' as postcolonial discourse: Towards a reassessment. *The Journal of Pacific History, 45*(3), 297–314.

Peters, K. (2020). The territories of governance: Unpacking the ontologies and geophilosophies of fixed to flexible ocean management, and beyond. *Philosophical Transactions of the Royal Society B, 375*(1814), 20190458. https://doi.org/10.1098/rstb.2019.0458

Ratuva, S. (2004). Reconceptualizing contemporary Pacific Islands states: Towards a syncretic approach. *The New Pacific Review, 2*(1), 246–262.

Remengesau Jr., T. E. (2016, September 1). *Our joint battle has just begun.* IUCN. Retrieved November 1, 2022, from https://2016congress.iucn.org/news/20160901/article/our-joint-battle-has-just-begun.html

Said, E. (1993). *Culture and imperialism.* Vintage.

Smyth, C., & Hanich, Q. A. (2019). *Large scale marine protected areas: Current status and consideration of socio-economic dimensions.* Pew Charitable Trusts. Retrieved September 1, 2023, from https://ro.uow.edu.au/lhapapers/3818/

Steinberg, P. E, & Peters, K. (2015). Wet ontologies, fluid spaces: Giving depth to volume through oceanic thinking. *Environment and Planning D: Society and Space, 33*(2), 247–264.

Steinberg, P. E. (2008). It's so easy being green: Overuse, underexposure, and the marine environmentalist consensus. *Geography Compass, 2*(6), 2080–2096.

Steinberg, P. E. (2009). Sovereignty, territory, and the mapping of mobility: A view from the outside. *Annals of the Association of American Geographers, 99*(3), 467–495.

Stokstad, E. (2023). Nations agree on long-sought high seas biodiversity treaty. *Science, 379*(6636), 971.

UK Foreign and Commonwealth Office. (2016, September 15). *Minister Duncan's Our Ocean Conference speech.* HM Government. Retrieved November 1, 2022, from https://www.gov.uk/government/speeches/minister-duncans-our-ocean-conference-speech--2

UK Foreign and Commonwealth Office. (2018, June 22). *Foreign Secretary announces UK strategy to protect world's oceans.* HM Government. Retrieved November 1, 2022, from https://www.gov.uk/government/news/foreign-secretary-announces-uk-strategy-to-protect-worlds-oceans

UK Foreign and Commonwealth Office. (2019, August 26). *Britannia protects the waves: £7 million extra funding to protect UK marine life.* HM Government. Retrieved November 1, 2022, from https://www.gov.uk/government/news/britannia-protects-the-waves-7m-extra-funding-to-protect-uk-marine-life

UNEP-WCMC & IUCN (2024). *Protected planet: The World Database on Protected Areas (WDPA) and world database on other effective area-based conservation measures (WD-OECM).* UNEP-WCMC and IUCN. Retrieved February 1, 2024, from www.protectedplanet.net

Can Borders in the Ocean Respond to Climate Change?

Yvonne Kunz

Introduction

Ocean health plays an extraordinarily important role in the current climate crisis (Franke et al., 2020). The oceans serve as a climate regulator by absorbing and redistributing carbon dioxide (CO_2) and heat. Oceans also support 'services' beyond climate regulations such as the provision of food (particularly protein), energy, health and well-being services, by offering a transport infrastructure, and by serving as cultural and recreational sites (IPCC, 2019). These functions will likely cease as the health of oceanic ecosystems deteriorates (von Schuckmann et al., 2021). For example, the oceans already seem to be losing their ability to act as carbon sinks (Craig, 2012; IOC-R, 2021) and coral reefs are dying from ocean warming resulting in complex cascading risks including the impact on the ocean's ability to provide food (Bruno et al., 2018). The

Y. Kunz (✉)
Sustainable Spatial Development and Governance, University of Trier, Trier, Germany
e-mail: kunzy@uni-trier.de

K. Peters and J. Turner (eds.), *Ocean Governance (Beyond) Borders*,
https://doi.org/10.1007/978-3-031-71322-4_7

biophysical relation between the ocean and the climate and accompanying consequences seem 'abundantly clear' (Holst & Rozemarijn, 2022). How to govern and manage these consequences is less clear.

The most prominent area-based global policy tool that is used to protect marine ecosystems is Marine Protected Areas (MPAs). MPAs, according to the globally authoritative definition of the International Union for the Conservation of Nature (IUCN), are 'clearly defined geographical spaces to achieve the long-term conservation of nature with associated ecosystem services and cultural values' (IUCN, 2015: n.p.). The IUCN (2017: n.p.) further states that MPAs are 'the only mainstream conservation-focused, area-based measure to increase the quality and extent of ocean protection. MPAs … offer nature-based solution[s] to support global efforts towards climate change adaptation and mitigation'. Yet, while oceans play a crucial role in sustainability and resilience, and MPAs are posited as *the* area-based tool to preserve oceans, the viability of MPAs is often questioned (Bates et al., 2019; Chuenpagdee et al., 2013; Di Franco et al., 2020; Gill et al., 2017). Controversies over the 'success' of MPAs span from parks that only exist on paper while not being actively managed (Grorud-Colvert et al., 2021), to ecosystems within protected areas deteriorating as fast as areas outside the protected space (Bates et al., 2019; Bruno et al., 2018). They also include ocean-grabbing discourses in which MPAs exclude communities from access to resources they depend on for their livelihood (Benjaminsen & Bryceson, 2012; Bennett et al., 2015).

These controversies only increase as climate change adds additional, external stressors into marine ecosystems (as well uses of the ocean being part of what feeds into the climate crisis). Surprisingly, however, little research has been conducted on the question of how MPA ecosystems are impacted by climate change, the role MPAs can play, and even less so on viewing this relationship critically (Roberts et al., 2017; Watson et al., 2014; Wilson et al., 2020). Meanwhile, at the United Nations Biodiversity Conference in December 2022, all members of the convention (196 in total) adopted the science-driven Kunming-Montreal Biodiversity Framework. One of its landmark agreements is the expansion of protected areas, on land and in the sea, to 30% by the year 2030—the so-called 30 × 30 agenda (Campaign for Nature, 2022; Gissi et al., 2022;

Sala et al., 2018; UNEP, 2022). Currently, depending on the definition and source, a share of approximately 8% of the oceans are under protection (Marine Conservation Institute, 2023), translating to the need for an immense increase of protected areas of 22% within less than seven years. MPAs, in this regard, are seen as a case of action to preserve biodiversity and support efforts towards climate change mitigation and adaptation.

If MPAs are to achieve their objectives of preserving marine ecosystems *and* creating more resilient biological communities that can resist and recover from climate events as the IUCN purports (Bates et al., 2019), they will need to consider climate change in their governance approach. *MPA governance itself has to adapt.* It is therefore essential to disentangle the markings and workings of MPAs and hence the logics that underwrite them (Peters, 2020). What constitutes them, how do they come into being and what are they capable of achieving? How are the borders between disciplines and the natural and social sciences also at play here? As MPAs are, ultimately, in the most fundamental sense about human-made and often contested compartments in the ocean, the chapter will focus on the lines constituting these so-called compartments that are vital in making MPAs. It does so to uncover how the MPAs in question were drawn, what informs the demarcation process and how MPAs affect human as well as non-human stakeholders in and adjacent to protected areas, and how human and non-human stakeholders engage with the conservation zones. It also crucially explores how these static lines are expected to take dynamic aspects like climate change into consideration. The chapter hence seeks to take a first step in critically reflecting the potential of the expansion of MPAs as part of the United Nations Biodiversity Framework towards marine biodiversity conservation in a changing climate.

In the following, this chapter will first conceptualise how using a border perspective can enhance a critical debate on MPA viability. To do so, terminology relevant for this chapter is anchored in the context of ocean governance, conservation practices and climate change. The methodology section introduces two case studies as well as the methods applied to learn from these sites. The ensuing part of the chapter discusses the discrepancy between attempt and achievement in border *marking* and border *making* along the two case study sites and how these sites incorporate (or not) climate change into their development. A reflection on conservation lines in the ocean and their potential contribution to halting biodiversity decline and climate change will close the chapter.

Conceptualising B/ordering Seascapes

'We live in a world of lines and compartments' (Newman, 2006: 1). These lines do not only organise national (terrestrial and marine) territories, countries and neighbourhoods; they are also used to govern and manage the use of natural resources. The idea of b/ordering environmental commons, such as oceans, is increasingly regarded as a strategy or a governance tool to protect fragile ecosystems from exploitation and overuse (Lambach, 2021; Miller, 2021). Although MPAs take an ecosystem approach, they are often focused on 'fixing' the issues related to the space they occupy. How they relate to the global spectre and reality of climate, which is both locally felt but clearly exists beyond borders, is a challenged.

While governance generally encompasses de facto as well as de jure regulations and practices, the focus of this chapter will be on de jure regulations, mainly on the formal management plans for MPAs to investigate how MPAs address climate change. Indeed, de jure regulations refer to the normative, usually legally enshrined, practices of governing. De facto regulations, the way the laws are perceived and translated into practice, often differ from normative law (Kunz et al., 2016). While focusing primarily on de jure regulations, the chapter acknowledges nevertheless, that 'rules enshrined in formal law [and management plans] provide only part of the picture' (Lund, 2008: 134). Similarly, there might be a discrepancy between attempt and achievement in (oceanic) biodiversity conservation that is rooted in (MPA) demarcation and delimitation processes.

Demarcation here is regarded as the process of deciding where socially constructed lines are drawn (Newman, 2006). In MPA development, these lines are supposed to be based on ecological features as much as possible. The lines drawn in this demarcation process make what the IUCN definition refers to as 'clearly defined geographical spaces' (Dudley, 2008: n.p.). The delimitation process, the way how boundaries are managed, can also differ from how the management plans of an MPA foresee it, when, for example, resources for enforcement are lacking. Demarcation (the marking of MPAs) and the delimitation (the working of MPAs) turn patches of the ocean into territories. Territories in the ocean (see also Peters et al., 2018) are then a political category, they can be 'owned, distributed, mapped, calculated, bordered and controlled' (Elden, 2010: 810). These territories are framed by borders understood as a 'practice of ordering and the discursive differentiation between us and them, seen through the lens of spatial bordering' (Van Houtum & Van Naerssen, 2002: 125).

At the World Congress of Protected Areas in Bali in 1982, it was noted that many marine protected areas, back then, were established as an extension of designated protected areas on land (International Union for Conservation of Nature & Natural Resources, 1982). The sea has long been regarded as wild, uninhabitable, uncontrollable place (Urbina, 2019), as 'wilderness external to human beings and separate from technology, culture, and daily life' (Elias, 2019: 15). While fishing practices of course existed, heavy resource exploitation as well as protecting ecosystems from exploitation has for long focused on terrestrial habitats. To this end, the potential tools for ocean governance have also been limited, with a prevalence of the simple transferring or copying of landed management into marine areas (Lambach, 2021; Peters, 2020). What might have been overlooked then and now, are 'wet ontologies', how the oceans exist as places with specific form, character and quality (Steinberg & Peters, 2015). Different from land, ocean governance needs to consider a temporal dimension pertaining to mobility of water and many of its inhabitants that is higher and less predictable than on its terrestrial counterpart (see also McAteer, volume). Further different from land, water's often liquid states permits modes of elemental immersion—a volume in distinction to grounded spaces (Steinberg & Peters, 2015).

MPAs, much like conservation areas more generally, are created through borders separating 'zonal areas of land or resource use' (Zimmerer, 2000: 361). Limiting the types of resource practices and containing those practices within geographically fixed areas (the aforementioned 'compartments'), they create common containers of resource use. These are at odds with the fluxes that are identified as a key feature of landscapes (Zimmerer, 2000) and even more so with the key features identifying seascapes. Furthermore, ocean, 'zones are established to reconcile conflicting activities in and around MPAs and (trying) to ensure that human activities do not negatively impact the biological features being protected' (Soemodinoto & Pedju, 2022: 63). In other words, this means that the demarcation processes are mimicked from processes on land, and supposedly based on biological features while they are also a political negotiation process.

Recent scholarly attention focuses on understanding borders and bordering as processes rather than focusing on material borders as simple *matters* of fact. According to Van Houtum and Van Naerssen (2002), b/ordering processes do not begin or end at a demarcation line, they are *negotiated* even before the boundary is drawn. Issues that borders seek to address are also often beyond them. Yet borders create exclusivity, with what seemingly belongs *inside* and what is not allowed to be part of the inside territory placed *outside* (see discussion, for example, in Johnson et al., 2011). In case of marine protected areas, there is a negotiating process on what is deemed worthy of being protected as positioned 'inside', vis-à-vis what is not worthy of being protected positioned on the outside. These negotiations are based on complex ecological but notably also political *values* concerning the environment. While real world negotiating processes and inherent power struggles are complex, we need to acknowledge this logic of MPA marking and working—the creation of insides and outsides—especially for those human and more-than-human actors not involved in bordering negotiations. How is global climate change considered while it exists both inside and outside the boundaries of MPAs—the bordered devices heralded as 'solutions' for it?

The cultural values driving MPA efforts also need to be considered. There is another demarcation issue at stake when considering MPAs and climate change: this is how disciplines work together. One challenge for successful conservation of habitats for the benefit of human and more-than-human actors might lie in crossing the separation between the natural and social sciences, yet another b/order or line of separation. While there is an agreement that complex real-world challenges like the preservation of biodiversity require interdisciplinary research and solutions (Freeth & Caniglia, 2020), social-ecological systems research engaging with climate change and sustainability themes hardly manages to equally bring the natural and the social sciences together (Biggs, 2021). The literature in the context of MPAs and climate change is natural science dominated. This chapter also seeks to contribute a first step towards balancing this out. The chapter contributes to better understanding current governance challenges of MPAs, especially under the additional stressor of climate change impacting humans and more-than-humans alike.

Methodology

Unpacking the working of MPAs towards a better understanding of MPA (border) challenges is focused on two case studies and a qualitative research approach, embedded into a transdisciplinary research project.

Research Sites

Caribbean waters are among the most biologically diverse marine regions in the world (Lotze, 2021). The Dutch Caribbean Waters are in the top five global biodiversity hotspots (IMARES Onderzoeksformatie, 2018). To allow an interregional comparison to better understand MPA governance practices and their potential for embracing climate change dynamics, two sites were selected that differ in age: Bonaire, one of the world's oldest MPAs; and Aruba, a recently established MPA. This selection offers insights into MPA governance from conservation areas initially established without considering climate change (Bonaire) in contrast to an MPA that actively intends to consider climate change from its inception (Aruba). The selection of sites was further informed by factors like size, design and accessibility.

The Bonaire National Marine Park[1] (BNMP) was established in 1979 and covers 2700 hectares (6700 acres) coral reef, seagrass and mangrove vegetation, as well as the sea around Bonaire and Klein Bonaire from the high-water line to a depth of sixty metres.

The lagoon Lac Bay (see Fig. 1, south-eastern shore) is also part of the underwater park and has been a Ramsar site since 1980. This means that the site is a unique wetland of international importance under the so-called Ramsar Convention, committed to a management that supports retaining its special character. In 1999, the area received the status of a national park from the Netherlands Antilles and the uninhabited island Klein Bonaire (see Fig. 1, western shore) was added as a legally protected nature reserve in 2001 (STINAPA, 2006). Management is in the hands of Stichting Nationale Parken Bonaire (STINAPA), a nature conservation non-governmental organisation (NGO) and a member of the Dutch Caribbean Nature Alliance (DCNA). The MPA runs around the whole island, with different use zones (marine reserve, no take zone, see Fig. 1). The whole coastline with its beaches is part of the protected area. Anchoring is not allowed anywhere around Bonaire, instead there is mooring infrastructure provided at which stopping of boats is permitted. Diving sites are clearly marked. In the so-called reserves diving is prohibited.

In 2016, the Government of Aruba, in partnership with the Netherlands Organization for applied scientific research TNO, was awarded a grant to establish an MPA on Aruba. This marked the beginning of a process establishing a marine park: 'Parke Marino Aruba was officially established by law AB 2018 no. 77 on 21 December 2018 and brought under the management of Fundacion Parke Nacional Aruba (FPNA) on 16 April 2019' (FPNA, 2019: n.p.). The marine park consists of four separate marine protected areas (see Fig. 2).

FPNA, the foundation in charge of managing the designated terrestrial protected area, was also appointed to manage the marine park (FPNA, 2019). The park aims to support the local economy by preserving the biodiversity, which would benefit fisheries as well as the tourism sector

[1] Marine (underwater) park and marine protected area are used interchangeability in this chapter.

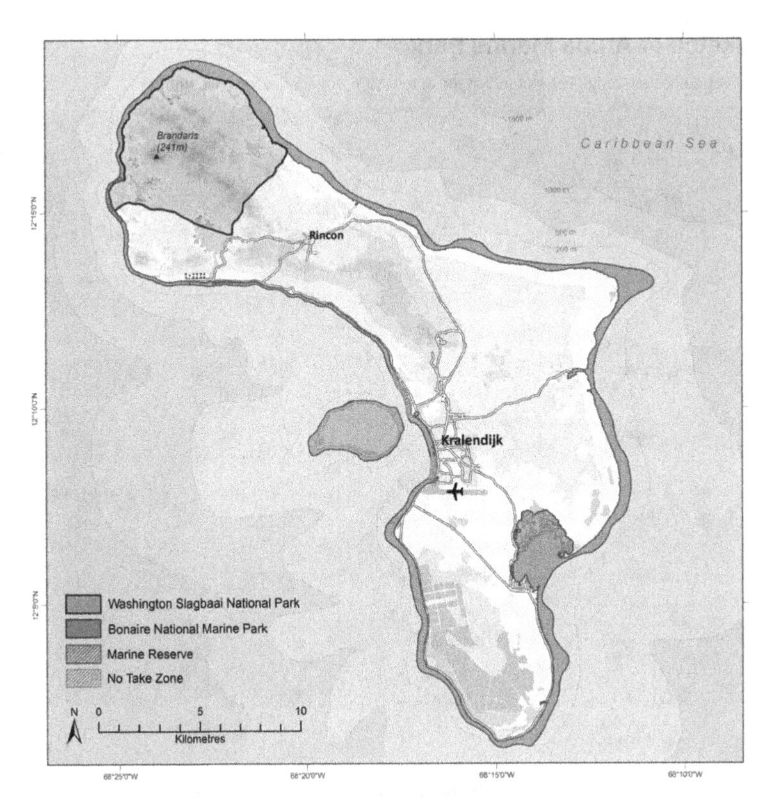

Fig. 1 Bonaire Marine Park. Source: STINAPA (2006)

(TNO, 2016). Unlike in Bonaire, where climate change was not explicitly mentioned, the first management plan, released in 2019, states that 'climate change adaptation activities will be of utmost importance during the two-year transition phase of the marine park' (FPNA, 2019: 5). In the first two years of its existence (2019–2021), the management plan of the park was tested [the period was extended]. It is referred to as transition period as the marine park has to be financially self-sustaining after two years and as it will undergo major evaluation and readjustment after four years, to be followed by an annual (financial) evaluation (ibid). This is a huge responsibility for the foundation as 'FPNA [the foundation] will manage Parke Marino Aruba, on behalf of the Aruban government, within the framework of Sustainable Development Goals

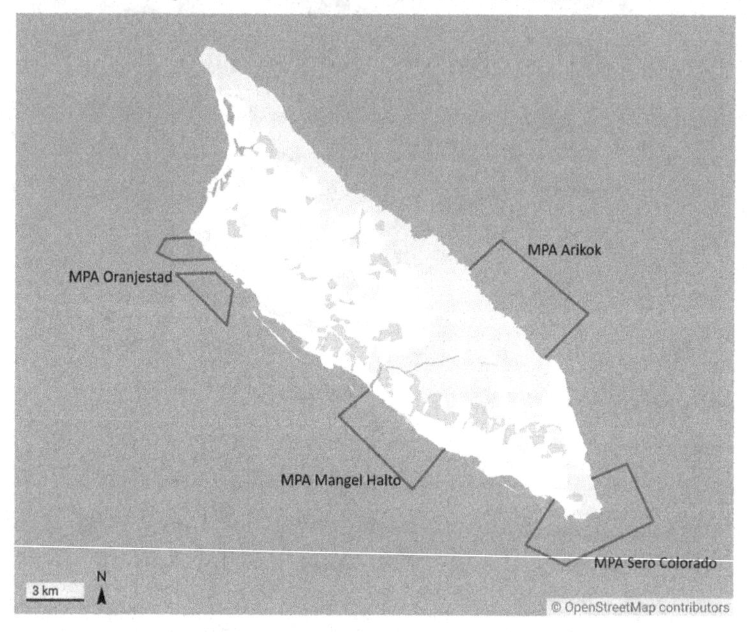

Fig. 2 Sketch of Aruba Marine Park. Produced by the author

(SDGs) and the Blue Economy concept and in accordance with national and international laws, treaties and conventions, standards and policies' (ibid).

Methods

In November 2021, a team consisting of a marine biologist, an aquatic ecologist, an anthropologist, a historian and a human geographer (author of this chapter) conducted a field stay, spending one week on Bonaire and Aruba, respectively. The primary focus of the field stay was to understand the viewpoints and practices of marine conservation practitioners towards MPA design, implementation and the role of climate change.

The field stay began by attending a DCNA workshop on Curacao, where numerous marine park managers from all Dutch Caribbean Islands gathered. A timeline and a Venn diagram were prepared with the group to understand when climate change entered the scene and to discuss stakeholders in the MPA context concerned with or working on climate change-related topics. Insights were then also obtained through participatory research methods on Bonaire and Aruba, and a number of semi-structured interviews, mainly with conservation practitioners, but also with government representatives, including the Aruban Tourism Authority, as well as fishery sector representatives. The interviews were recorded, transcribed and processed through a qualitative content analysis, shedding light on the marking and making of the MPA and its considerations for climate change. Of interest in this analysis were perceptions on how the MPA design was informed, what it attempts and what it is expected to achieve. The management plans of the respective marine parks, their websites and complementary documents were also part of this analysis.

Compartmentalising the Sea in Bonaire and Aruba

The two case studies introduced will guide a discussion of the marking and making of Marine Protected Areas on the islands of Bonaire and Aruba to generate insights that might help to better understand MPAs and their place in the context of the Kunming-Montreal Biodiversity Framework in a changing climate.

Border Marking: Drawing Lines in the Sea

As lines in the ocean are what constitutes marine protected areas, a closer look at what informs these lines, what they attempt and what they can achieve is a necessary starting point. On Bonaire, the MPA surrounds the island of Bonaire and Klein Bonaire (see Fig. 1). The boundaries of the marine park were, according to the management document, marked to be easily identifiable (STINAPA, 2006). The landward border is the high-water mark, meaning that 'anything which is wet during the course of the normal tide' (STINAPA, 2006: 80) is included. The seaward border is the 60 metre depth mark (translating horizontally), is less visible than the high-water mark, which is visible, or in effect drawn, through a debris line as well as on the mangrove prop roots. Taking the three-dimensionality of the marine environment into account, the management plan also marks out the potential of protecting the seabed. Anchoring is prohibited within the marine park, while mooring devices have been installed by STINAPA to be used by anyone on a boat. There are additional delineations within the MPA, separating two marine reserve areas where neither diving nor fishing is allowed (Fig. 1). According to STINAPA (2006), these reserves were not respected from 1983 to 1990 when the park was not actively managed.

The reserves are reference sites to monitor how areas *not* impacted by water sports develop. While the eastern shore is very rough and not suitable for diving, unless for very experienced divers, the western shore is peppered by dive sites. Due to the steep drop-off close to the shore, shore diving is the usual choice, where divers and snorkelers just walk into the water instead of having to take a boat to dive sites. To avoid destruction by trampling vast areas when entering the waters, yellow stones labelled with entry/exit serve as indicators for divers and snorkelers, at least when approaching the dive sites from the shore. This is another form of border marking/making. This still needs to be enforced by STINAPA as some tourists also disregard these points. As one interview partner told us: '… but the people just think: I want to go here. … Let me go to a place where there is nobody' (26.11.1).[2]

The two marine reserves are thus marked by signboards or lines on land: those constructed on the map and those also (simultaneously) on the shore. A challenge is that 'while the boundaries are clearly visible on [this] map, the boundaries are not physically visible [when on or in the water] and are prone to be disregarded by fisheries and other resource users' (Mac Donald, 2022: 160). What is marked on paper, in short, is

---

[2] These numbers represent our own interview, internal coding system to ensure anonymity of interlocutors.

not marked on the sea. Moreover, and pertinent to this chapter, while we frequently heard about the need to adapt to climate change *on the land*, to rising sea levels for example, resulting in flooding, or to massive amounts of sargassum arriving at the shores of Bonaire, these aspects are not (or cannot be) mirrored in the management plan and the marking of the MPA.

The Aruba National Park Foundation (ANPF), in charge of managing the Marine Protected Areas, would be in favour of an all island-round MPA, similar to the design of the MPA on Bonaire: 'Our ... opinion was that it should be an island-round marine park. ... and almost everyone said it should be an island-round marine park' (23.11.1). There are, however, four MPAs, of which one is split into two to allow the cruise ships to pass through. This creates a lot of more lines, constructing borders, compared to Bonaire. Towards the land, the border again is the shore. Towards the sea, the border lies at two kilometres into the sea. It was not possible to find out during the interviews conducted in the field, how the location of the four MPAs was decided on apart from learning that the design has been established through TNO, the Dutch organisation for applied research.

Whether climate change has been considered in the design remains an open question. Aruba National Park Foundation (AFPN), mandated to manage the MPAs, is currently in the process of designing zones within these four MPAs through a stakeholder consultation process. The result of this will be different use zones and further line drawing and the marking of more borders. Yet as one anecdote that was shared with us regarding the consultation sessions and climate change revealed, climate change remains elusive:

> ... What is really funny or really interesting is that ... after day 1 we did a little recap of what we did that day, positive things, negative things, so that we might adjust some things if needed for the next day. One of the people in the group mentioned 'I'm missing climate change'. So she really said like 'we have been talking about all of these things, but we didn't discuss climate change at all ...'. Everybody agreed. 'Yes! It's missing from what we have mentioned.' (23.11.1)

On Aruba, signboards on land inform people about protected areas and also on the objective of the protected areas. ANPF is not equipped with a boat yet so that controlling (or enforcing) the borders of the plan, most of which lie in the waters, are still challenging endeavours. Also, there is a challenging financial dimension: 'There are a lot of borders. Marine conservation has a lot of costs, both buoy maintenance, all of these things' (23.11.1).

Moreover, all of these borders, whether on Bonaire or Aruba, are meant to restrict human activities, to avoid harm caused by people. They are designed to serve non-humans by ideally altering human behaviour. STINAPA (2006: 27) outlines in their management plan that the lines, the borders, are to protect 'planktonic and pelagic sea creatures including fish and migratory species such as dolphins and turtles, coral reefs, sea grass beds and including surface dwelling animals and plants and infauna'. Similarly, ANPF aims at protecting significant nesting sites for sea turtles and marine birds, shark nurseries, seagrass, seahorse and shark habitats as well as diverse coral reefs. However, neither the fish nor temperatures, nutrients, sargassum (macro algae) nor other small or big organisms stop at these borders. Migratory fish or fish migrating because of *changes in conditions such as temperature*—climate change—cannot be stopped by these boundaries. They are mere markings on paper, with the ocean resisting such inscription (see also Lehman et al., 2021). While the Bonaire MPA management plan speaks about locations of certain ecosystems, such as seagrass beds and coral reefs, these locations and the decision on where to place the marine reserve are not brought together in a transparent way in the management plan. And yet, even if the Bonaire management plan does not reveal information on how choices were made, there is still a major difference between both sites. The Bonaire delineations do reflect a much greater ambition with regard to the protection of an entire ecosystems, rather than seemingly random bits and pieces as is the case in Aruba. Aruba more openly reflects the real-world negotiations of demarcation processes as the four patches seem to the outcome of a negotiation process. Even if we lack information on the processes, the borders do share information on the politics behind both plans and what was possible at what time.

Border Working: Discrepancy of Threats and Conservation Potential

In an attempt to understand what boundaries are capable of preventing, stopping, compartmentalising, or in a broader sense, what borders are capable of ordering (and in view of the climate crisis), I juxtapose goals of the MPAs and threats to the marine ecosystems as described in the management plans of the MPAs and the role MPA borders can play in achieving these goals by preventing or diminishing threats.

The Bonaire management plan (STINAPA, 2006: 52) explains that 'understanding the goals and objectives of a protected area as well as the range of direct and indirect threats are essential elements of any management plan'. The items listed as main threats are disturbances caused by SCUBA divers, yachting, fishing activities and coastal development, nutrient enrichment, sedimentation, landscaping practices, building and constructing in the coastal zone. The management plan also acknowledges that threats 'may originate from within the protected area or from beyond its boundaries' (2006: 82), recognising that 'significant barrier[s] to conservation' emerge from 'the considerable links between the terrestrial and marine environment' given that 'everything which happens on the land directly and immediately impacts on the island's fragile coral reefs' (STINAPA, 2006: 82). This reveals the dilemma of the MPA, given that most of the threats lie outside of its borders and that the implementation of a line does little to deter these threats. Climate change as a global process, locally felt, exceeds the lines but is less acknowledged than the specific human uses (which may of course be part of it) but which in turn makes climate appear more elusive.

This assessment has been reconfirmed by a more recent iteration of STINAPA—the so-called Conservation Action Plan—where again more of the threats identified are located outside than inside the borders of the marine protected area. Here climate change does feature prominently in the Action Plan. One out of four strategic areas is a 'Climate Change Readiness Strategy for Nature Conservation on Bonaire'. The document is an internal working document that has to my knowledge not been followed up on. Part of the strategy would have been to 'understand impacts, … identify and prioritize potential partners to help STINAPA address climate change on Bonaire … and to develop climate change adaptation/resilience recommendations' (internal document). This may be because it is unclear how to achieve these goals, but also because there are often more pressing issues to attend to like sargassum washing in unhealthy amounts to the coastal ecosystem arriving with a need for direct action.

Moreover, overfishing is a challenge that the MPA borders do attend to by prohibiting fishing activities in certain areas, the monitoring of success remains challenging. 'On Bonaire, the catches and market prices are not monitored or regulated' (Mac Donald, 2022: 168) indicating that there is no systematic monitoring of what is being taken out by the fishery sector. This makes it difficult to assess what harm the fishery sector does and it seems challenging to assess what impact the marine reserve has on halting negative impacts caused by overfishing. At the same time, observations of stakeholders shared with us during the interviews mirror Mac Donald's (2022) findings that the fishery sector does not seem to be the major challenge. As one interview partner put it: 'the fishery sector is not the biggest pressure we have, even though we want a more sustainable fishery sector' (26.11.2). And again, the study by Mac Donald reveals that 'there is no (long term) data available regarding Bonaire's catch landing (2022: 178). While it is acknowledged that fisheries might be a problem, to a certain level, compared to other threats, and particularly those threats outside the border of the MPA, including climate change, the damage caused by few commercial fishers active on the island is not the most pressing threat.

A similar discrepancy between main threats identified and potential working of borders can be found in the Aruba Preliminary Management Plan (FPNA, 2019). The major challenges to successfully achieving self-set objectives are sustainable financing, lack of awareness and data deficiency. While these can be overcome through lobby work of staff members, the threats following next on the list are harder to be addressed by MPA delimitation: coastal development, marine and coastal recreation (unregulated), land, air and marine pollution, maritime activities, invasive species and diseases, wildlife feeding, extractive industries, climate change and natural disasters (ibid). The majority of the threats identified by the management plan actually lie outside the borders of the

MPAs and hence outside the scope of what the staff members and the lines have power over, including consequences of climate change. Additional to the borders not being able to protect the protected area from major identified threats, a recently published study reveals that while the 'recently assigned marine parks generally includes some of the reefs of highest natural values on Aruba, … the MPA Oranjestad, especially in the northern section, does not harbor reef communities of significant value' (CARMABI, 2020: 56). The document further states that stretches in between two MPAs, hence not protected areas between MPA Mangel Halto and MPA Oranjestad (see Fig. 2), 'harbor some of the best remaining reef along Aruba's leeward shore' (CARMABI, 2020: 56).

On both islands, stakeholders reported a lack of scientific data to actually make precise statements on what is at risk. 'The island is very data deficient, so that's a real big problem, because we need that for proper management decisions' (29.11.1) is what one interview partner explained to us in this regard. Both management plans state that overfishing is a threat that the MPA can halt if implemented and managed efficiently while data to confirm this statement seems to be lacking. The already mentioned document on the 'Coral Reef Baseline Study for Aruba 2019' actually starts by stating that 'there is almost no systematic information about the state of the marine ecosystem in Aruba' (CARMABI, 2020: 9). While the borders are supposed to be scientifically informed to support the conservation and sustainable use of resources, this does not seem to be the case in our case study sites. This makes the borders less meaningful, for humans and non-humans alike and then in turn less convincing too, as a strategy for climate change.

Conclusion

Recent literature urges us to think about climate change and biodiversity loss as being intimately intertwined (Pörtner et al., 2021; Shin et al., 2022). MPA governance needs to embrace climate change. This has become clear from the literature, from the interviews as well as from the management and action plans on the islands. While climate change as a threat to marine ecosystem is prevalent on the islands, how to manage the sea in the face of the threats of it remains a question yet to be answered. MPAs are positioned as part of the 'solution' because they govern activities linked to climate change but those links are not clear and hence the current challenges of MPA governance need to be addressed while taking climate change consequences into the equation. Pushing for an expansion of protected areas in the sea while not addressing current challenges seems to end in what Fletcher (2023) labels as 'failing forward'.

Acknowledging in this chapter the de jure practices of MPA development it is possible to see that MPA border marking and making are also power laden processes. The knowledge considered in the demarcation process, the border marking, is not evenly distributed across stakeholders. Recognition might open up space for a more diverse set of knowledges to be considered, which could in turn allow for more flexible approaches to ocean governance. Presenting a design by an external organisation to a local foundation managing the park to then conduct a stakeholder consultation seems at odds with the potential knowledge local resource users could contribute to the process (see also Dawson et al., 2021). As an earlier quote noted, 'I'm missing climate change'. Listening to knowledge about the impacts and effects of climate change might be one way to imagine and implement it into plans, even as plans as site-specific tools in the face of an issue that is certainly felt within but which also exceeds any line-drawing exercise.

While Aruba and many other recently established MPAs do take note of consequences of climate change, they are yet to get a grip on how to consider these consequences in their governance approach (and to question whether that approach can be a 'solution'). As a learning process, future research should look into discrepancies of conservation practices on the ground and management plan regulations as climate change might be addressed differently in daily practices, more than an inert management plan can reveal. What needs to be avoided is that MPA practitioners are left with the feeling 'of not having the luxury to deal with climate change' while being overwhelmed by other, seemingly more pressing challenges even as those challenges may be climate related.

The Kunming-Montreal Biodiversity Framework seems noble in its intention, but seems to lack a reality check on the ground. Simply expanding the surface area under protection, does not help to achieve the expected positive outcomes. Bordering more of the ocean will not necessarily equate to mitigating or reversing climate impacts. Indeed, if MPAs are struggling to handle current challenges, this will not be solved by an increase in area. MPA governance is overwhelmed in attempts to achieve their objectives: power struggles, finances and threats originating from outside its border and hence outside its control. The lines that will be drawn under the umbrella of the latest Biodiversity Framework will remain as static and powerless towards threats as they have been under earlier campaigns.

A mere expansion and increase of static borders will not be able to address dynamic challenges of climate change. What seems necessary from the research here that is climate change is sometimes (not always) acknowledged, and where it is, it is unclear how MPAs can remedy, measure or monitor or bring change in response. More scientific data, involvement of local knowledge from the very beginning is needed, before (more) lines are drawn in the sea. Integrated thinking considering the space where threats emerge as well as where they hit is also vital.

All these themes are disturbed by lines that we need to break through. In case scientific knowledge is informing the MPA design, it is too often only a natural science perspective that is not able to grasp the whole picture, especially not with power imbalances at play. There is also a line separating the international (or national) level expertise, from local knowledge. If local knowledge, the situated expertise adapting to climate change on an everyday basis and where a global crisis is *felt,* is not considered as valuable as international level decisions, it will remain extremely difficult to find contextualised responses better able to respond to contextualised threats. This also means that the solutions will remain abstract, and detached from the local reality on the ground.

A further line that we need to break through is the line between the place of threat origin and the place of threat impact. We need to consider coastal MPAs in their coastal condition. This applies to a landscape perspective as well as to the acknowledgement of coastal communities living with the land and the sea, making them experts of this integrated system perspective. There is a need to find a way to protect marine ecosystems, from climate change as well as from other threats (be they climate linked or not). The push for 30% of the oceans under protection by 2030 appears, after what this chapter has been presenting, as a rather ad hoc campaign that desperately feels the need to act, without halting to assess what kind of action would be viable, especially in times of climate crisis.

References

Bates, A. E., Cooke, R. S. C., Duncan, M. I., Edgar, G. J., Bruno, J. F., Benedetti-Cecchi, L., & Côté, I. M. (2019). Climate resilience in marine protected areas and the 'protection paradox.' *Biological Conservation, 236*(August), 305–314.

Benjaminsen, T. A., & Bryceson, I. (2012). Conservation, green/blue grabbing and accumulation by dispossession in Tanzania. *Journal of Peasant Studies, 39*(2), 335–355.

Bennett, N. J., Govan, H., & Satterfield, T. (2015). Ocean grabbing. *Marine Policy, 57*, 61–68.

Biggs, R. (Ed.). (2021). *The Routledge handbook of research methods for social-ecological systems*. Routledge International Handbooks.

Bruno, J. F., Bates, A. E., Cacciapaglia, C., Pike, E. P., Amstrup, S. C., van Hooidonk, R., Henson, S. A., & Aronson, R. B. (2018). Climate change threatens the world's marine protected areas. *Nature Climate Change, 8*(6), 499–503.

Campaign for Nature. (2022). *More than 100 countries commit to protect at least 30% of land and oceans by 2030.* Retrieved May 5, 2024, from https://www.campaignfornature.org/more-than-100-countries-commit-to-protect-at-least-30-of-land-and-oceans-by-2030#:~:text=More%20than%20100%20Countries%20Commit,by%202030%20%E2%80%94%20Campaign%20For%20Nature

Caribbean Research and Management of Biodiversity Foundation (CARMABI). (2020). *Coral Reefs Baseline Study for Aruba 2019.* Retrived November 11, 2024, from https://www.dnmaruba.org/wp-content/uploads/2023/08/Aruba-Coral Reef-Report_2021-compressed.pdf

Chuenpagdee, R., Pascual-Fernández, J. J., Szeliánszky, E., Alegret, J. L., Fraga, J., & Jentoft, S. (2013). Marine protected areas: Re-thinking their inception. *Marine Policy, 39*, 234–240.

Craig, R. K. (2012). Ocean governance for the 21st century: Making marine zoning climate change adaptable. *Harvard Environmental Law Review, 36*(2), 305–350.

Dawson, N. M., Coolsaet, B., Sterling, E. J., Loveridge, R., Gross-Camp, N. D., Wongbusarakum, S., Sangha, K. K., Scherl, L. M., Phuong Phan, H., Zafra-Calvo, N., Lavey, W. G., Byakagaba, P., Idrobo, C. J., Chenet, A., Bennett, N. J., Mansourian, S., & Rosado-May, F. J. (2021). The role of Indigenous peoples and local communities in effective and equitable conservation. *Ecology and Society, 26*(3), 19. https://doi.org/10.5751/ES-12625-260319

Di Franco, A., Hogg, K. E., Calò, A., Bennett, N. J., Sévin-Allouet, M.-A., Alaminos, O. E., & Lang, M. (2020). Improving marine protected area governance through collaboration and co-production. *Journal of Environmental Management, 269*(September), 110757. https://doi.org/10.1016/j.jenvman.2020.110757

Dudley, N. (2008). *Guidelines for applying protected area management categories.* IUCN.

Elias, A. (2019). *Coral empire: Underwater oceans, colonial tropics.* Duke University Press.

Elden, S. (2010). Land, terrain, territory. *Progress in Human Geography, 34*(6), 799–817.

Fletcher, R. (2023). *Failing forward: The rise and fall of neoliberal conservation.* University of California Press.

Franke, A., Blenckner, T., Duarte, C. M., Ott, K., Fleming, L. E., Antia, A., & Reusch, T. B. H. (2020). Operationalizing ocean health: Toward integrated research on ocean health and recovery to achieve ocean sustainability. *One Earth, 2*(6), 557–565.

Freeth, R., & Caniglia, G. (2020). Learning to collaborate while collaborating: Advancing interdisciplinary sustainability research. *Sustainability Science, 15*(1), 247–261.

Fundacion Parke Nacional Aruba (FPNA). (2019). *Parke Marino Aruba. Preliminary management plan 2019–2021.* Retrieved from May 30, 2024, from https://www.dcbd.nl/sites/default/files/documents/PNA_Management-Plan-20192021_Digital_compressed-1.pdf

Gill, D. A., Mascia, M. B., Ahmadia, G. N., Glew, L., Lester, S. E., Barnes, M., & Craigie, I. (2017). Capacity shortfalls hinder the performance of marine protected areas globally. *Nature, 543*, 665–669.

Gissi, E., Maes, F., Kyriazi, Z., Ruiz-Frau, A., Santos, C. F., Neumann, B., & Quintela, A. (2022). Contributions of marine area-based management tools

to the UN Sustainable Development Goals. *Journal of Cleaner Production, 330*, 129910. https://doi.org/10.1016/j.jclepro.2021.129910
Grorud-Colvert, K., Sullivan-Stack, J., Roberts, C., Constant, V., Horta e Costa, B., Pike, E. P., Kingston, N., … & Lubchenco, J. (2021). The MPA guide: A framework to achieve global goals for the ocean. *Science, 373*(6560). https://doi.org/10.1126/science.abf0861
Holst, R., & Rozemarijn, J. (2022). The Climate-Oceans Nexus. In P. G. Harris (Ed.), *Routledge handbook of marine governance and global environmental change* (pp. 28–38). Routledge.
IMARES Onderzoeksformatie. (2018). *Staat van de natuur van Caribisch Nederland 2017: een eerste beoordeling van de staat (van instandhouding), bedreigingen en managementimplicaties van habitats en soorten in Caribisch Nederland.* Wageningen Marine Research.
Intergovernmental Panel on Climate Change (IPCC). (2019). Summary for policymakers. In H.-O. Pörtner, D.C. Roberts, V. Masson-Delmotte, P. Zhai, M. Tignor, E. Poloczanska, K. Mintenbeck, … & Weyer, N. M. (Eds.), *IPCC special report on the ocean and cryosphere in a changing climate.* Retrieved May 16, 2024, from https://www.ipcc.ch/srocc/chapter/summary-for-policymakers/
International Union for Conservation of Nature (IUCN). (2015). *Protected areas as tools for disaster risk reduction. A handbook for practitioners.* Retrieved May 16, 2024, from https://portals.iucn.org/library/efiles/documents/2015-001.pdf
International Union for Conservation of Nature (IUCN). (2017). *Marine protected areas and climate change. International Union for the Conservation of Nature.* Retrieved May, 26, 2024, from https://www.iucn.org/resources/issues-briefs/marine-protected-areas-and-climate-change
International Union for Conservation of Nature and Natural Resources. (1982). *The World National Park Congress.* Retrieved May 16, 2024, from https://portals.iucn.org/library/sites/library/files/documents/1982-005.pdf
Johnson, C., Jones, R., Paasi, A., Amoore, L., Mountz, A., Salter, M., & Rumford, C. (2011). Interventions on rethinking 'the border' in border studies. *Political Geography, 30*(2), 61–69.
Kunz, Y., Hein, J., Mardiana, R., & Faust, H. (2016). Mimicry of the legal: Translating de jure land formalization processes into de facto local action in Jambi Province, Sumatra. *Austrian Journal of South-East Asian Studies, 9*(1), 127–146.
Lambach, D. (2021). The functional territorialization of the high seas. *Marine Policy, 130*, 104579. https://doi.org/10.1016/j.marpol.2021.104579

Lehman, J., Steinberg, P. E., & Johnson, E. R. (2021). Turbulent waters in three parts. *Theory & Event, 24*(1), 192–219.

Lotze, H. K. (2021). Marine biodiversity conservation. *Current Biology, 31*(19), R1190–R1195.

Lund, C. (2008). *Local politics and the dynamics of property in Africa.* Cambridge University Press.

Mac Donald, S. (2022). *Life in "Paradise". A social psychological and anthropological study of nature conservation in the Caribbean Netherlands* (Doctoral Dissertation), Leiden University.

Marine Conservation Institute. (2023). *The marine protection atlas.* Retrieved May 5, 2024, from https://mpatlas.org/

Miller, M. A. (2021). B/Ordering the environmental commons. *Progress in Human Geography, 45*(3), 473–491.

Newman, D. (2006). The lines that continue to separate us: Borders in our 'borderless' world. *Progress in Human Geography, 30*(2), 143–161.

Peters, K. (2020). The territories of governance: Unpacking the ontologies and geophilosophies of fixed to flexible ocean management, and beyond. *Philosophical Transactions of the Royal Society B, 375*, 20190458. https://doi.org/10.1098/rstb.2019.0458

Peters, K., Steinberg, P. E., & Stratford, E. (Eds.). (2018). *Territory beyond terra.* Rowman & Littlefield.

Pörtner, H. O., Scholes, R. J., Agard, J., Archer, E., Arneth, A., Bai, X., … & Ngo, H. (2021). Scientific outcome of the IPBES-IPCC co-sponsored workshop on biodiversity and climate change. Retrieved May, 26, 2024, from https://doi.org/10.5281/ZENODO.4659158

Roberts, C. M., O'Leary, B. C., McCauley, D. J., Cury, P.M., Duarte, C. M., Lubchenco, J., Pauly, D., … & Carlos Castilla, J. (2017). Marine reserves can mitigate and promote adaptation to climate change. *Proceedings of the National Academy of Sciences, 114*(24), 6167–6175.

Sala, E., Lubchenco, J., Grorud-Colvert, K., Novelli, C., Roberts, C., & Rashid Sumaila, U. (2018). Assessing real progress towards effective ocean protection. *Marine Policy, 91*, 11–13.

Shin, Y.-J., Midgley, G. F., Archer, E. R. M., Arneth, A., Barnes, D. K. A., Chan, L., & Smith, P. (2022). Actions to halt biodiversity loss generally benefit the climate. *Global Change Biology, 28*(9), 2846–2874.

Soemodinoto, A., & Pedju, M. (2022). Evaluability assessment of Indonesian marine conservation areas for management effectiveness evaluation. *ILMU KELAUTAN: Indonesian Journal of Marine Sciences, 27*(1), 61–72.

Steinberg, P. E., & Peters, K. (2015). Wet ontologies, fluid spaces: Giving depth to volume through oceanic thinking. *Environment and Planning D: Society and Space, 33*(2), 247–264.

Stichting Nationale Parken Bonaire (STINAPA). (2006). *Bonaire National Marine Park Management Plan 2006*. Retrieved May 5, 2024, from https://discomap.eea.europa.eu/map/Data/Milieu/OURCOAST_052_NL/OURCOAST_052_NL_Doc4_Bonaire.pdf

The Integrated Ocean Carbon Research (IOC-R). (2021). *Integrated Ocean Carbon Research: A summary of ocean carbon research, and vision of coordinated ocean carbon research and observations for the next decade*. UNESCO

United Nations Environment Programme (UNEP). (2022). *UN biodiversity conference (COP 15)*. Retrieved May 5, 2024, from http://www.unep.org/un-biodiversity-conference-cop-15

Urbina, I. (2019). *Outlaw sea. Die gesetzlose See: ein Augenzeugenbericht*. Topicus.

van Houtum, H., & Naerssen, T. V. (2002). Bordering, ordering and othering. *Tijdschrift Voor Economische En Sociale Geografie, 93*(2), 125–136.

von Schuckmann, K., Le Traon, P.-Y., Smith, N., Pascual, A., Djavidnia, S., Gattuso, J.-P., Grégoire, M., … & Zupa, W. (2021). Copernicus marine service ocean state report, issue 5. *Journal of Operational Oceanography, 14*(sup1), 1–185.

Watson, J. E. M., Dudley, N., Segan, D. B., & Hockings, M. (2014). The performance and potential of protected areas. *Nature, 515*(7525), 67–73.

Wilson, K. L., Tittensor, D. P., Worm, B., & Lotze, H. K. (2020). Incorporating climate change adaptation into marine protected area planning. *Global Change Biology, 26*(6), 3251–3267.

Zimmerer, K. S. (2000). The reworking of conservation geographies: Nonequilibrium landscapes and nature-society hybrids. *Annals of the Association of American Geographers, 90*(2), 356–369.

Bordering Marine Belonging: The Meanings, Mobilities and Materialities of Bioinvasion

Satya Savitzky, Kimberley Peters, and Katherine G. Sammler

Introduction

So-called invasive or 'non-native species' pose what many scientists deem to be a significant threat to global biodiversity (Barker & Francis, 2021), with some noting that 'invasive alien species are one of the *leading* threats to the biodiversity around the world' (Khan et al., 2022: 459, emphasis added). The use of the word 'threat' undoubtedly produces a narrative regarding the impacts of invasive species, a phenomenon that became 'observable' to naturalists 'as far back as the 18th Century', the point at which it would be constructed and defined as a feature of the natural world also by the *leading* scientists of the age, such as Augustin Pyramus

S. Savitzky · K. Peters (✉)
Marine Governance, Helmholtz Institute for Functional Marine Biodiversity at the University of Oldenburg, Oldenburg, Germany
e-mail: kimberley.peters@hifmb.de

K. G. Sammler
University of Twente, Enschede, Netherlands
e-mail: k.g.sammler@utwente.nl

K. Peters and J. Turner (eds.), *Ocean Governance (Beyond) Borders*,
https://doi.org/10.1007/978-3-031-71322-4_8

de Candolle and Charles Darwin (Gentili et al., 2021: 1). Indeed, invasive species were understood to be 'species outside of their bioregion which can possibly *threaten* native ones (De Candolle, 1855; Darwin, 1859; Pyšek et al. 2004)' (Gentili et al., 2021: 1, emphasis added). This definition holds, with the International Union for the Conservation of Nature (IUCN) currently describing invasive species as 'animals, plants or other organisms that are introduced by humans, either intentionally or accidentally, into places outside of their natural range, negatively impacting native biodiversity, ecosystem services or human economy and well-being' (IUCN, n.d.: n.p.).

Indeed, the IUCN and many narratives of what is otherwise called 'bioinvasion' often stress the negative impacts of such 'introductions' of species from one place to another—a movement that sees species cross borders from 'here' to 'there' via a number of human acts: a 'combination of colonial rule, rapid economic development and increasing global trade exchanges across continents and borders' (Gentili et al., 2021: 1). Whilst the damaging or 'deleterious' impacts of invasive species are well noted (ibid) and often relate to global efforts of management due to their role in global biodiversity loss and its associated ills (including in the United Nations Sustainable Development Goal 15, see IUCN, n.d.), bioinvasion raises important questions of belonging within the natural world and the movement of life around our planet. As Inglis writes,

> Animals which have been labelled 'invasive species' are the great villains of the wildlife conservation world. They are represented, both by the popular media and within academic discourse, as marauders, aliens, killers and monsters (Strayer and Waldman 2013). As a result, the public is encouraged to perceive these animals, not as valuable members of the biotic community, but as a threat that needs to be met with deadly force. (Inglis, 2020: 299)

Whilst Inglis notes that the introduction of 'invasives' '*can* cause harm to other species', they also urge that 'regardless of the validity of the above premise, the invasive species narrative is fundamentally flawed', not least on moral grounds, and there is a need to change how we speak of invasive species (2020: 300, original emphasis).

This chapter builds from this premise to focus specifically on invasive species—particularly more-than-human marine life and aquatic species—and to consider how questions of what belongs where in the ocean realm are shaped by meanings of 'nativeness'; the movement of things from 'here' to 'there' (or where they are understood to belong, or not); and the materialities that may be entangled with such meanings and mobilities (i.e. vectors of invasion, such as 'species *on* (as fouling communities), *in* (as boring communities [in old wooden vessels]) and *inside* (as ballast communities)' (Ojaveer et al., 2018: 2). It further seeks to consider how borders are enacted insofar as bio-risks, in order to contain them. It aims to continue the critical work of social science and humanities scholars exploring biogeography and bioinvasion (from Robbins [2004] to Barker and Francis [2021]) to consider how such phenomena are constructed and how borders, mobility and belonging are central to the framings of the 'problem'. It is not the intention of the chapter to, like Inglis (2020), comment on whether invasives can or do result in 'harm', and where, how and to what. Rather, this chapter has two aims. It firstly identifies invasive species as a topical issue that remains marginal in critical discussions of *ocean* bordering/belonging. Secondly, it puts forward three ways in which bioinvasion, bordering and belonging can be better understood at sea, through (1) questions of definition and meaning; (2) an attention to the mobilities and materialities enrolled within processes of bioinvasion—the movement of marine species in global circuits; and (3) how borders are enacted in view of invasive species via biosecurity measures. The chapter seeks to disrupt conventional narratives of marine belonging, showing marine lifeworlds that are rather more entangled than spatial, bio- and ecological divisions would suggest.[1]

[1] Since this chapter was written, a new paper has emerged on the 'terminological tempest of invasion science' (Soto et al., 2024). Although making different arguments to our chapter with its focus on belonging, it is a relevant piece with which readers may wish to also engage.

Ocean Borders, Belonging and Bioinvasion: A Topic on the Margins?

The topic of bioinvasion, as noted in the introduction, is one that rests upon ideas of *where* specific forms of (more-than-human) life *belong* or not. As Mee and Wright note, '[b]elonging connects matter to place, through various practices of boundary making and inhabitation which signal that a particular collection of objects, animals, plants, germs, people, practices, performances, or ideas is meant "to be" in a place' (Mee & Wright, 2009: 772). Belonging, then, is a *matter* that rests on defining borders. It rests on what a defined (marine) space—its physical, biological and ecological make-up—*should* look like in its 'natural' composition (and at a specific time). Consequently, when species are 'out of place' they are deemed, through dominant discourse, to transgress. If they trangress, they do not belong (Cresswell, 1992, 2014) and come to hold negative associations, which, in turn, produce 'exclusionary processes' (Mee & Wright, 2009: 772) or, in the case of invasive species, exclusionary governance (where management techniques often call for removal and 'restoration'). Alderman (2018) notes that such narratives around invasive species that rest on concepts of bordering and belonging fail to account for the more complex entanglements between more-than-human life and its spatial contexts. As he writes, 'invasive organisms move, develop materially, and take on meaning' through their *relations* (Alderman, 2018: 6). In other words, the simplistic bordering narratives around bioinvasion produce the binaries of belonging or not (which in turn relate the development of polarising terminology such as 'negative impacts', and 'threats').

Much of this narrative is centred within various natural science work and economic analysis that has been conducted on the 'problem' (see, e.g., Marbuah et al., 2014), whereby invasives have been posited to have had a 'minimum economic cost' of 'USD 1.288 trillion' between 1970 and 2017 (IUCN, n.d.: n.p.). Whilst there is critical work in the natural sciences that questions the linguistic as well as biophysical basis of understandings of bioinvasion (see Inglis, 2020), notions of what should be where—notions of belonging—hold fast. So too does a historical emphasis on landed cases of bioinvasion. Indeed, despite the topic's inherently transdisciplinary nature-cross-cutting socio-cultural, economic, political and environmental concerns, social science literature on invasive species is now much more marine in focus, but has, in the past, historically been landlocked.

Emphasis on biodiversity loss has focused on the terrestrial realm, perhaps because here the impacts of bioinvasion are more readily seen and felt (in landscape change and knock-on effects in crop production, etc.). As Bax et al. put it: 'a massive invasion by a marine species is often much less conspicuous than, for example, an invasion by a brightly flowering garden plant' (2003: 317). On land, bioinvasion may also offer more accessible and feasible research opportunities where the very materiality of 'earth' means it may be more easily divided up into ecosystems with clear boundaries (in spite of complexities around elevation, or dominant vegetation type, which are used as a distinguishing feature for defining ecotone boundaries). As Ojaveer et al. notes (2018: 12) '[m]ajor research focus on marine invasions is relatively recent, emerging initially in the 1960s and 1970s in a few regions, such as the Panama Canal, Suez Canal, and the Pacific coast of North America. … As a result NIS [non-indigenous species] data varies considerably among geographic regions and taxonomic groups, resulting in significant imbalance among marine taxa in inventories'. The relative dearth of oceanic knowledge means it is harder to say, more easily, what 'belongs' where in the ocean because ocean borders are enacted on maps but are trickier to define in the mobile, churning, voluminous space of the seas themselves (see Steinberg & Peters, 2015). Yet, in spite of this, marine invasive species have been documented across all ocean basins, including in the polar regions (Miller & Ruiz, 2014), with introduced species ranging from fish and seaweeds to microorganisms.

Where watery spaces have featured in invasives research, work has been on inland waterways. However, even this 'wet' focus was to come rather later to the study of invasions as established by ecologist Charles Elton, the so-called father of invasive science (Gentili et al., 2021: 1). Indeed, Ricciardi and MacIsaac (2011: 212) note that 'the ecological impacts of freshwater invasions were rarely studied until many years after the publication of Elton's (1958) influential book', *The Ecology of Invasions by Animals and Plants*. Water bodies beyond—the seas and oceans—would come later still and remain rather more on the margins (reflecting also their Western ontological positioning, see Steinberg, 1999, 2001 and Peters, 2010). Indeed, '[w]hile the global transport of ballast water (and associated sediments) was first recognized as a potential dispersal mechanism for plankton in the late 1890s, quantitative research on the issue does not appear in the primary scientific literature until the mid-1980s' (Bailey et al., 2005: 261). Ballast refers to any heavy material placed low in a vessel (in a tank) to provide stability when crossing the tumultuous ocean. Water (and all that is contained within it, including marine life) is often used as ballast—which, taken in at one port is released at the next as cargo loads change. This makes and remakes ocean socio-ecologies.

Yet it is really only in the more recent past that the natural sciences now feature an increasing body of work that does focus on the ocean 'invaders' that move with ocean currents, the conditions of climate change and as stowaways in the holds of ships. Techniques developed to understand marine invasion vary, from ecosystem analysis, to the modelling of ship movements to illustrate the global trends of invasives as linked to global trade. Some natural science research has also attended to marine invasive species management. Yet, in the social sciences and humanities, discussions remain sidelined. Indeed, in the *Routledge Handbook of Biosecurity and Invasive Species* (Barker & Francis, 2021)—a comprehensive 20-chapter guide to the topic—only a handful of chapters have a saltwater focus (see Guo on islands [2021] and Giesler and Cottier-Cook [2021] on marine and coastal systems), reflecting the terrestrial dominance of research. This may relate to the aforementioned issues of 'easier' access to research sites on land and/or long-cited issues afflicting marine and maritime research by social science and humanities scholars pertaining to technology, skill, expertise and cost (see Peters, 2010; Psuty et al., 2004). One might surmise that this more broadly reflects the general land-centric biases of the social sciences and humanities before the oceanic turn (Anderson & Peters, 2014; DeLoughrey, 2019; Peters, 2010; Steinberg, 2001). Together with the marginal positioning of the oceans, questions of oceanic belonging and marine life in invasives studies have been largely bordered-off (see exceptions, David & Gollasch, 2017; De Lucia, 2019).

In what follows we bring these debates to the fore through the perspective of mobilities studies, (more-than) human geographies and the critical (marine) social sciences; three interlocking fields where debates on bioinvasions circulate, but where the seas and oceans remain on the margins. Bringing these interests and fields together, this chapter raises the necessity and potential of examining *marine* belonging through interrogating the political 'work' of definitions (of native/non-native/invasive), followed by investigating articulations of marine invasion via routes and infrastructures (ships, tanks and the materialities of invasion). It finally addresses debates concerning biosurveillance and security. We turn to these sections next beginning with a discussion of meaning-making around borders, belonging and bioinvasion.

On Definition and Meaning: Spatial Bounding and Categorising Life

There is no doubt that invasion science responds to a very real problem (see also Inglis, 2020). The world's merchant fleet transports every kind of cargo and all variety of aquatic species, unintentionally circulated via the human current of ballast water exchange (David & Gollasch, 2017). The Ballast Water Management Convention (BWM Convention) began in 2004 under the longer title the 'International Convention for the Control and Management of Ships' Ballast Water and Sediments' to counter the 'threat' of bioinvasion. This international regulation, introduced by the International Maritime Organisation (IMO), came into force in 2017 as a governance mechanism to manage cross-border movements of unwelcome aquatic life within a highly mobile and 'fast' industry. It is only since 2017 that ships moving over 400 gross tonnes must now legally comply with a set of measures including having a vessel ballast management plan, log book and certificate of compliance by the flag state (see IMO, 2022). This regulatory 'landscape' is driven by the overwhelming consensus that 'invasive alien species' (so defined) can be ecologically and economically damaging (Molnar et al., 2008; Zenni et al., 2021) and that border management (through plans, log books and certification) can be enacted to prevent 'threats'.

However, the binary between 'native' and 'non-native' at the core of invasion science, governance and conservation practice has been disputed with increasing ferocity in recent decades, as a problem encompassing ethical and normative issues (Inglis, 2020). The prevailing invasion science paradigm rests on ideas that every species has a specific geography to which it 'belongs' (Warren, 2021). This distribution of species is studied in biogeography, the field of biology that deals with the geographical distribution of plants and animals as spatial and temporal bounding of already-categorised life. Categorising species as either 'native' or 'alien', with natives embraced and aliens rejected, is central to mainstream invasion biology and nature conservation practice. It is determined by understandings of habitat boundaries and 'natural' geographical range. Yet, it has become apparent that 'the question of what is native and what is alien is riddled with paradox and puzzle' (Smout, 2014: 12).

Precise definitions of 'native' and 'alien' are elusive but, as commonly understood, native species are those that evolved in an area or colonised it 'naturally'. In contrast, alien species are those that have been introduced beyond their natural range by humans, whether intentionally or otherwise (Warren, 2021: 19, see also Gentili et al., 2021). Native animals and plants are deemed to 'belong' in an area, alien ones are 'out of place' (Gibbs et al., 2015). Yet, no species is intrinsically alien or native but is deemed so only in relation to a particular area at a particular time, such that the spatial and temporal boundaries of that space can be—and are—constructed in many different ways.

For instance, the temporal threshold of nativeness varies widely between nations, with the date of European colonisation often selected as a convenient but arbitrary 'year zero' (Warren, 2021: 3). This is a problematic baseline that erases the agency of Indigenous peoples in changing the landscape—despite the vast evidence to the contrary. Robbins and Moore argue that some so-called Edenic sciences are preoccupied with 'what constitutes a "natural" system state, and over what kinds of interventions, if any, should be advocated by scientists' (2012: 3), seeking to return to baselines imagined as 'untouched', 'pure', and 'pristine' nature, before the 'fall'. The nativeness of species is often defined in relation to political territories (which have legal responsibilities for protecting biodiversity), even as many national borders often cross-cut ecological units (Warren, 2021: 7). As such, socially constructed borders are at odds with the ways nature moves and 'belongs'. Consequently, many critics argue that 'the terms [such as "native" or "non native"] are sufficiently arbitrary, malleable and contingent that the labels make little ecological sense' (Warren, 2021: 3). Yet the continued use of the terms has material impacts in projects seeking eradication of 'invaders'. There are also ideological implications. The historical development of this language (e.g. in ecological literature) resonates with xenophobic, anti-immigration, nativist and military language in Western countries (Larson, 2005; Lower & Sturtevant, 2021; Tsing, 1995). Crucially, ideas of nativeness and alienness have emerged concomitant with the 'nationalization of nature and the naturalization of nation' (Warren, 2021: 8), with consequent conflations of ecological and political nationalisms (Antonsich, 2021).

What is more, the 'native vs. alien' conceptual framework (and the wilderness management practices derived from it) are argued to be insufficient in facing the novel dynamics of the Anthropocene (Hill & Hadly, 2018). Invasion science tends to presume that a homology exists between geographic spaces and environmental variables to provide an objective approach to describe invasive species. Yet, as points in geographic space are divorced from their historic environmental variables due to rates of environmental change, place-based conceptions of 'nativeness' become increasingly redundant. One infamous terrestrial example is the expectation that in a warmer, drier future the range of Joshua Trees will have shifted poleward and to higher elevations, to the extent that may no longer be extant in their eponymous national park (Sweet et al., 2019). This has resulted in debates over the principles of assisted migration as, 'historical baselines for ecological protection and restoration have become outdated ... [yet] "managed relocation" remains ecologically uncertain and, for some, ethically questionable' (Lekan, 2022: 3). In particular, it is thought that climate change 'will progressively warp habitats until the species lose the collection of environmental parameters to which they were once native': put more simply, 'no species will be truly "native" in the Anthropocene' (Hill & Hadley, 2018: n.p.). For marine spaces and aquatic habitats, the issue of historical species range is even more fraught given the challenges of knowing, bounding and mapping such variables. Baselines for many marine ecosystems are generally even less well known and shift perceptually (Hobday, 2012). However, the extreme mobility possible during the lifecycle of aquatic flora and fauna, along with somewhat limited knowledge bases, pose distinct questions to the concepts outlined in this chapter: 'native', 'non-native' and 'alien'.

The Mobilities, Materialities and the Matter of 'Invasion'

The problematic basis of the previously described definitions also calls attention to the relevance of mobilities for questions of belonging and bordering. Non-human organisms have long been moved across vastly separated localities and ostensible ecological and anthropological borders, but 'invasion biologists'—the name given to the branch of ecologists who study the impacts of 'introduced' species—argue that rates of travel increased dramatically from the twentieth century onwards (Pimentel, 2002, see also Gentili et al., 2021). This is due to the proliferation of global routes—of intensified mobilities—during this time, most significantly aviation and shipping (Lewis & Maslin, 2015). In responding to Barua's recent call (2021: 1469) for interdisciplinary scholarship that brings together studies of sociotechnical systems, ecology and *mobilities* research, we argue that it is only through such a lens that we can hope to understand the relationship between species, dynamism and space, and trouble the enduring narratives concerning borders of belonging in species location.

Despite, then, their inherent capacity for movement via this global infrastructure, aquatic invasive species have been surprisingly underexamined within the burgeoning social science literature on oceans and shipping (see David & Gollasch, 2017). The 'mobilities' literature on shipping (i.e. Birtchnell et al., 2015; Cresswell & Martin, 2012) pays scant attention to invasive species (although the 'parasitic' harnessing of shipping infrastructure has been a topic, see Martin [2013]). Such literature has instead focused on logistics (e.g. Birtchnell et al., 2015), labour relations (e.g. Borovnik, 2012), the economic and political impacts of containerisation (e.g. Heins, 2016), the geopolitical importance of shipping routes and so on (Savitzky, 2016). We might say that the focus has been overwhelmingly on the significance of shipping for *human-centred* globalisation, stitching together societies-and-economies into a global—yet heterogeneous and differentiated—consumer culture. Mobilities literature remains primarily focused on the question of what these mobilities are for people—consumers, producers, seafarers, shipping lines and so on (e.g. Hulme [2020]).

Yet, shipping (and to a lesser extent aviation) is increasingly associated with the production of other kinds of globalisation: the globalisation of the world's biota, which will leave a far greater imprint than any of the consumer ephemera delivered in shipping containers. *Infrastructure*—road networks, shipping routes, railways, aviation, etc.—has become a popular topic in the social sciences over the last few decades (i.e. Anand, 2017; Graham & Marvin, 2002; Harvey & Knox, 2015; Larkin, 2013; Martin, 2016; Urry, 2007). Despite this work's putative commitment to recognising the differences made by other-than-human forces, it remains anthropocentric, 'seldom registering' according to Barua (2021: 1468), 'the role of infrastructure in affording (or preventing) the mobility of other-than-human life'. Yet, as more and more species are moved around the globe via shipping, the planet becomes ever closer to resembling its state 300 million years ago, when all land on Earth was lumped together in a single supercontinent.

Indeed, as Anthropocene theorists Lewis and Maslin write (2018: 102), 'in the sixteenth century [the point at which transcontinental shipping networks were established] a new planet-wide human-driven evolutionary experiment began. … What plate tectonics did over tens of millions of years is being undone by shipping in a few centuries and aviation in a few decades. We are creating a New Pangea'. Such a New Pangea is meant to represent how the oceanic barriers that prevented many terrestrial species from easily moving between continents were erased through the global mobility of modern transportation infrastructures. The singular ocean surrounding the Pangea supercontinent has been designated Panthalassa. The authors make no mention of the infrastructural creation of a New Panthalassa, although that is arguably the marine analogy of this planet-wide human-driven evolutionary experiment. The result of upending 300 million years of evolutionary place-based adaptations in a comparatively short amount of time is throwing species together into new ecological juxtapositions and has, as yet, unquantifiable detrimental impacts for biodiversity.

Yet, the global fleet of over 50,000 ships, which Barua (2021) usefully characterised as 'mobile infrastructure', have come to 'dictate non-human movement at unprecedented scales and speed'. According to the Global Biodiversity Outlook 5, the flagship publication of the Convention on Biological Diversity, the 'projected growth in global shipping traffic is likely to increase the risk of alien species invasions by between three and 20 times the current level by 2050' (Secretariat of the Convention on Biological Diversity, 2020: 140). Research has shown that ship mobilities are a primary biological invasion *vector* (Ruiz & Carlton, 2003). There are several (sub)vectors of ships that move species around, including ballast water (Carlton, 1985) and ballast sediment (Bailey et al., 2005), biofouling (Ashton et al., 2016), internal tank biofouling (Drake et al., 2005) and internal seawater piping systems (Lewis & Dimas, 2007). Cargo is another vector through which marine organisms propagate, although cargo is more often associated with the unintended transfer of terrestrial species—demonstrating the complex interlinkages between the marine and terrestrial. These subvectors highlight how the materialities of ships and shipping create mobility pathways for biological media, eliminating the traditional borders or 'barriers' isolating species and habitats. Accordingly, mobilities—from shipping, to the species that ships (and tanks) move and the very movement of continents—are pivotal to thinking about the fundamental question of marine belonging. However, these movements—and the infrastructures that sometimes underscore them (such as tanks)—also matter and have matter.

Indeed, the shipping container's significance is by-now well analysed; an invention without which economic globalisation would be unthinkable (Martin, 2016). Nevertheless, the ballast tank as a stubbornly tangible, material entity that is fundamental to the operation of container ships, has received very little attention. As previously mentioned, the ballast tank is a compartment within a boat, ship or other floating structure that holds water and provides stability to vessels and maintains a vessel's seaworthiness. Up until the mid-twentieth century, sea-going vessels used solid ballast to steady the vessel. This comprised sandbags, rocks or iron blocks, which were loaded/unloaded once the cargo loading process was finished. However, today's vessels carry liquid ballast, which includes freshwater, saltwater or brackish water in various ballast tanks. Modern transoceanic shipping moves several billion tonnes of ballast water annually, accounting for the transport of 10,000 species each day (Streftaris et al., 2005). Bulkers and tankers are the dominant movers of ballast water (Davidson et al., 2018). It is ballast that matters, then, to the movement of species from 'here' to 'there'. The ballast tank is the vector from which discharged species may then be deemed to belong, or not (David & Gollasch, 2017).

Recent work has shown how the documented history of marine invasions coincides with the period of diversification of ships (Davidson et al., 2018). Species transfers increased with expanding world trade, resulting from, first, the increasing number and size of vessels, second, the increasing speed of vessels and third, increasing global port connectivity (Davidson et al., 2018). The specialisation or diversification of the global shipping fleet has also affected opportunities for species transfers, based upon changes in ship structure and 'behaviour', though as Davidson et al. (2018) write, and as we continue to echo, this has received relatively little attention. What is moved are not just individual animals but entire aquatic assemblages, 'intensifying both the volume and taxonomic diversity of biotic flows and transforming the very composition of aquatic ecosystems' (Barua, 2021: 1472). Ominously, introduced species sometimes remain innocuous for decades, then begin to spread and produce impacts; impacts of some introduced species are not recognised quickly or, crucially, are enhanced after another species is introduced, such that some introduced species that are currently apparently harmless may nonetheless warrant concern. This uncertainty—due to species cumulative or non-linear interactions—gives rise to pre-emptive governance logics, as critical social science scholars have observed (though still largely of terrestrial accounts, see especially Barker and Francis [2021] and Dobson et al. [2013]). In other words, bordering devices are enacted, in advance, to try and control movements 'here' to 'there' through circuits of global trade. Often these devices are materially manifested in tank technologies and also in performed practices of testing tank 'matter' for its composition of (micro)organisms.

As countries attempt to manage risks of bioinvasion, *biosecurity* approaches have been adopted and adapted from one governing context to the next. A clear example is the appropriation of border technologies for managing humans in the context of controlling more-than-human biologies, or the practices of terrestrial borders adapted for coastal and marine enforcement. The pre-border domain is increasingly significant here, as states attempt to shift the risk of biosecurity off-shore (Barker, 2015). Such techniques produce new regimes of off-shore biosurveillance and biosecurity. Indeed, in our final section, we ask how borders are enacted in view of invasive species via biosecurity measures.

Security, Surveillance and Pre-emptive Biogovernance at the Border

Common biosecurity practices per se include pre-checking goods and passengers in departure ports, developing import requirements, screening import applications, developing risk methodologies for pathways, and pre-departure education and communication programmes (Dobson et al., 2013: 12–13). A state may have bilateral agreements with numerous import countries determining pre-border quarantine measures. In addition to formal agreements, bilateral exchanges of information, practices and advice are also significant. A characteristic feature of biosecurity is that it can refer 'simultaneously to the mundane and the extraordinary', from precautions such as hand-washing and disinfection, to complex surveillance webs of biocontrol (Barker & Francis, 2021: 1).

Biosurveillance, defined as: 'the production, analysis and circulation of information on potential invasive events or epidemics', is a crucial ongoing aspect of biosecurity practice (Dobson et al., 2013: 12). This term encompasses forms of surveillance that monitor the health and status of plants, animals and ecosystems or progression of a virus during an epidemic. These practices involve 'mixtures of humans, nonhumans and technologies performing different practices, including visual inspections, counting, photographing, reporting, sniffing, X-raying, measuring, swabbing, weighing, scanning, recording, collecting and sampling'—practices through which biological markers are rendered as information (Dobson et al., 2013: 13). For shipping, and ballast particularly, there are processes for tank management that include disinfection, and rules on dumping/discharging. There is also 'pathway surveillance', which targets high-risk sites to look for organisms that may be entering across a nation's borders. Governance works by acting to regulate those pathways (Dobson et al., 2013).

Marine species introductions are influenced, as noted in the previous section, by multiple factors including: 'ship type, voyage duration, ballast water uptake volume and location, discharge volume and location, environmental differences between the source and destination ports, and the environmental tolerance of the organisms' (Saebi et al., 2020: 2). Assessing species introduction and dispersal risks is a complex task involving many types of data from many different sources. Vector interception or disruption has been identified as 'the most vulnerable and directly manageable portion of the invasion sequence' (Carlton & Ruiz, 2005: 37). Governance practices often encode so-called species 'watch lists'—that is, species deemed to present an 'immediate or potential' threat to economy, environment or human health. According to the United States National Environmental Policy Act of 1970, the term 'invasive species' ('alternatively known as an alien, exotic, injurious, introduced or naturalized, non-native, nonindigenous, nuisance, or noxious species') means a 'non-native' species whose introduction 'does or *is likely*' to cause economic or environmental harm or harm to human health (Johnson et al., 2017: 1). Of concern therefore are not only the *actual* but also the *potential* impacts of species. Reiterating our discussion on definitions and meanings, this makes categorisation as 'non-native' enough for a species to be considered dangerous—despite the fact that only a small number of introduced species have gone on to become 'invasive' and associated with widespread economic damage. As such, bordering is also pre-emptive—processes of exclusion happen prior to any 'threat'. Such trends mirror the violent trends also linked to strategies of human bordering in some nation states, where potential or 'future' risks and mobile pathways are identified, and then locked down. Indeed, once identified, 'alien' species are targeted for destruction, in biopolitical practices that 'in the same gesture protects life and produces death' (De Lucia, 2019: 4). Here protecting the life of that which 'belongs', sustains that belonging through the border expulsion of the 'other'.

Conclusions

The economic significance of global cargo shipping is by now commonly acknowledged (Birtchnell et al., 2015). Shipping moves more than 90% of global trade (George, 2013) from the shoes on our feet, to computers we use and the food we eat (Peters, 2010). However, this is not all that the world's container ships and tankers transport. Whilst today the deliberate transport of animals, such as cattle, still occurs, shipping fleets shape the unintentional movement of other more-than-human biological life at the microscopic, although arguably far more prolific, scale (Helmreich, 2009). As shipping traverses the globe in new circuits of supply and demand, ballast water is used to counterbalance the delivery of cargo by appropriately 'weighting' vessels as they move to their next destination. Accordingly, water is 'picked up' in one part of the world and 'dropped off' in another (Seebens et al., 2013), moving vast numbers of organisms with the water and with the ship. Species are 'displaced' and purportedly 'invade' new spaces at unprecedented scales and speeds.

This chapter has posited the need to further explore bioinvasion in (salt) watery spaces and through a specifically social science and humanities lens. The social scientific work on invasive species that does exist has focused overwhelmingly on land, leaving an oceanic lacuna. In contrast, this chapter has gone to sea to pose how we might better understand the mobilities of invasive species, governance and security practices, and how 'invasibility' and 'nativeness' are understood by various scientists and practitioners. It has done so to unpack the fundamental question that lies at the heart of understandings of invasion: what belongs where—and how borders function in this narrative as both ideological constructions for justifying management and also practical interventions in subsequent techniques of governance and control at sea.

To this end, we have addressed two aims: the first centred on shifting socio-political studies of marine invasive species from the 'margins to the centre of academic vision' (Lambert et al., 2006: 488) under the corpus of work constituting the 'oceanic turn' that is focused on critical discussions of ocean bordering/belonging. To date much work on borders, bordering and belonging is human focused, with a large and important emphasis on migration politics (see also Şarlak, this volume). But more-than-human life also matters, not least with bioinvasion identified as a global problem both economically and environmentally. Second then, we have moved through three ways in which bioinvasion, bordering and belonging can be better understood, exploring definitional work, the role of mobilities and materiality and finally the enacting of ocean governance via biosecurity measures.

These are just a few lines of enquiry. Indeed, in opening up space to explore biogovernance in this chapter, other avenues to make sense of these highly contentious issues might arise, linked to how further complex governance arrangements are enacted at sea; or how the fluid materiality of the oceans affects the planning, enactment and ratification of legal instruments. All of this is necessary because,

> The ocean has a paradoxical character. It simultaneously separates and connects. It is, or was, the great barrier dividing people and places, but has increasingly come to serve as the great facilitator of all sorts of cross-border exchanges, as diverse as consumer objects and invasive species.... (Savitzky, 2016: 81)

Understanding the 'paradoxical character' of the ocean means being attentive to the dynamics of separation and connection—'here' and 'there', 'belonging' and 'exclusion'—and the ways borders write the oceans even though the ocean's fluidity often works to defy inscription. This chapter has pushed for a thinking of/about marine belonging, identifying the studies of invasion so far evaded. It has put forward three ways for better understanding bioinvasion in the oceanic realm, acknowledging land and sea not as discrete, but as world of 'cross-border' exchanges, of spaces that are ultimately 'stitched' together (Savitzky, 2016).

Acknowledgements We thank Dr Sandip Hazareesingh for his continual encouragement, support and kindness in the process of completing this chapter.

References

Alderman, D. H. (2018). Toward a historical geography of human–invasive species relations. In G. L. Buckley, & Y. Yolonda Youngs (Eds.), *The American environment revisited: Environmental historical geographies of the United States* (pp. 3–18). Rowman and Littlefield.

Anand, N. (2017). *Hydraulic city: Water and the infrastructures of citizenship in Mumbai.* Duke University Press.

Anderson, J., & Peters, K. (2014). *Water worlds: Human geographies of the ocean.* Farnham.

Antonsich, M. (2021). Natives and aliens: Who and what belongs in nature and in the nation? *Area, 53*(2), 303–310.

Ashton, G. V., Davidson, I. C., Geller, J., & Ruiz, G. M. (2016). Disentangling the biogeography of ship biofouling: Barnacles in the Northeast Pacific. *Global Ecology and Biogeography, 25*(6), 739–750.

Bailey, S. A., Nandakumar, K., Duggan, I. C., Van Overdijk, C. D., Johengen, T. H., Reid, D. F., & MacIsaac, H. J. (2005). In situ hatching of invertebrate diapausing eggs from ships' ballast sediment. *Diversity and Distributions, 11*(5), 453–460.

Barker, K. (2015). Biosecurity: Securing circulations from the microbe to the macrocosm. *The Geographical Journal, 181*(4), 357–365.

Barker, K., & Francis, R. A. (Eds.). (2021). *Routledge handbook of biosecurity and invasive species.* Routledge.

Barua, M. (2021). Infrastructure and non-human life: A wider ontology. *Progress in Human Geography, 45*(6), 1467–1489.

Bax, N., Williamson, A., Aguero, M., Gonzalez, E., & Geeves, W. (2003). Marine invasive alien species: A threat to global biodiversity. *Marine Policy, 27*(4), 313–323.

Birtchnell, T., Savitzky, S., & Urry, J. (2015). *Cargomobilities: Moving materials in a global age.* Routledge.

Borovnik, M. (2012). The mobilities, immobilities and moorings of work-life on cargo ships. *Sites: A Journal of Social Anthropology and Cultural Studies, 9*(1), 59–82.

Carlton, J. T. (1985). Transoceanic and interoceanic dispersal of coastal marine organisms: The biology of ballast. *Oceanography and Marine Biology: An Annual Review, 23*, 313–371.

Carlton, J. T., & Ruiz, G. M. (2005). Vector science and integrated vector management in bioinvasion ecology: Conceptual frameworks. In H. A. Mooney, R. Mack, J. A. McNeely, L. E. Neville, P. J. Schei, & J. K. Waage (Eds.), *Invasive alien species: A new synthesis.* (pp. 6–58). Island Press.

Cresswell, T. (1992). *In place/out of place: Geography*. University of Minnesota Press.

Cresswell, T. (2014). *Place: An introduction* (2nd edition). Wiley-Blackwell.

Cresswell, T., & Martin, C. (2012). On turbulence: Entanglements of disorder and order on a Devon beach. *Tijdschrift Voor Economische En Sociale Geografie, 103*, 516–529.

David, M., & Gollasch, S. (2017). Ballast water and harmful aquatic organism mobilities. In J. Monois (Ed.), *Maritime Mobilities* (pp. 11–137). Routledge.

Davidson, I. C., Scianni, C., Minton, M. S., & Ruiz, G. M. (2018). A history of ship specialization and consequences for marine invasions, management and policy. *Journal of Applied Ecology, 55*(4), 1799–1811.

DeLoughrey, E. (2019). Toward a critical ocean studies for the Anthropocene. *English Language Notes, 57*(1), 21–36.

De Lucia, V. (2019). Bare nature. The biopolitical logic of the international regulation of invasive alien species. *Journal of Environmental Law, 31*(1), 109–134.

Dobson, A., Barker, K., & Taylor, S. L. (Eds.). (2013). *Biosecurity: The socio-politics of invasive species and infectious diseases.* Routledge/Taylor & Francis Group.

Drake, L. A., Meyer, A. E., Forsberg, R. L., Baier, R. E., Doblin, M. A., Heinemann, S., Johnson, W. P., Koch, M., Rublee, P. A., & Dobbs, F. C. (2005). Potential invasion of microorganisms and pathogens via 'interior hull fouling': Biofilms inside ballast water tanks. *Biological Invasions, 7*(6), 969–982.

Elton, C. S. (1958). *The ecology of invasions by animals and plants.* Methuen.

Gentili, R., Schaffner, U., Martinoli, A., & Citterio, S. (2021). Invasive alien species and biodiversity: Impacts and management. *Biodiversity, 22*(1–2), 1–3.

George, R. (2013). *Deep sea and foreign going: Inside shipping, the invisible industry that brings you 90% of everything*. Portobello Books.

Gibbs, L., Atchison, J., & Macfarlane, I. (2015). Camel country: Assemblage, belonging and scale in invasive species geographies. *Geoforum, 58*, 56–67.

Giesler, R. J., & Cottier-Cook, E. J. (2021). Marine and coastal ecosystems. In K. Barker & R. Francis (Eds.), *Routledge handbook of biosecurity and invasive species* (pp. 142–160). Routledge.

Graham, S., & Marvin, S. (2002). *Splintering urbanism: Networked infrastructures.* Routledge.

Guo, Q. (2021). Island ecosystems. In K. Barker & R. Francis (Eds.), *Routledge handbook of biosecurity and invasive species* (pp. 128–141). Routledge.

Harvey, P., & Knox, H. (2015). *Roads: An anthropology of infrastructure and expertise.* Cornell University Press.

Heins, M. (2016). *The globalization of American infrastructure: The shipping container and freight transportation.* Routledge.

Helmreich, S. (2009). *Alien ocean: Anthropological voyages in microbial seas.* University of California Press.

Hill, A. P., & Hadly, E. A. (2018). Rethinking "native" in the Anthropocene. *Frontiers in Earth Science, 6*(96), 1–4.

Hobday, A. J. (2012). Review of shifting baselines: The past and the future of ocean fisheries, edited by J.B.C. Jackson, K. Alexander, and E. Sala. *Oceanography, 25*(1), 306–308.

Hulme, A. (2020). *On the commodity trail: The journey of a bargain store product from east to west* (Illustrated Edition). Routledge.

Inglis, M. I. (2020). Wildlife ethics and practice: Why we need to change the way we talk about 'invasive species.' *Journal of Agricultural and Environmental Ethics, 33*(2), 299–313.

International Maritime Organization (IMO). (2022). *Implementing the Ballast Water Management Convention.* Retrieved December 20, 2022, from https://www.imo.org/en/MediaCentre/HotTopics/Pages/Implementing-the-BWM-Convention.aspx

International Union for Conservation of Nature (IUCN). (n.d.). *Invasive alien species.* Retrieved May 1, 2024, from https://iucn.org/our-work/topic/invasive-alien-species#:~:text=Invasive%20alien%20species%20are%20animals,human%20economy%20and%20well%2Dbeing

Johnson, R., Crafton, R. E., & Upton, H. F. (2017). *Invasive species: Major laws and the role of selected federal agencies.* Congressional Research Service. Retrieved May 1, 2024, from https://crsreports.congress.gov/product/pdf/R/R43258/7

Khan, R., Iqbal, I. M., Ullah, A., Ullah, Z., & Khan, S. M. (2022). Invasive alien species: An emerging challenge for the biodiversity. In M. A. Öztürk., V. Altay., & R. Efe (Eds.), *Biodiversity, conservation and sustainability in Asia:*

Volume 2: Prospects and challenges in South and Middle Asia (pp. 459–471). Springer International Publishing.

Lambert, D., Martins, L., & Ogborn, M. (2006). Currents, visions and voyages: Historical geographies of the sea. *Journal of Historical Geography, 32*, 479–493.

Larkin, B. (2013). The politics and poetics of infrastructure. *Annual Review of Anthropology, 42*, 327–343.

Larson, B. M. H. (2005). The war of the roses: Demilitarizing invasion biology. *Frontiers in Ecology and the Environment, 3*, 495–500.

Lekan, T. M. (2022). An otherworldly species: Joshua trees and the conservation-climate dilemma. *The Rachel Carson Center Review, 1*. https://doi.org/10.5282/rcc-springs-1471

Lewis, J. A., & Dimas, J. (2007). *Treatment of biofouling in internal seawater systems-Phase 2*. Defense Science and Technology Organization Victoria (Australia) Maritime Platforms Division. Retrieved May 24, 2024, from https://collection.sl.nsw.gov.au/record/74VvKBzAw7E3

Lewis, S. L., & Maslin, M. A. (2015). Defining the anthropocene. *Nature, 519*(7542), 171–180.

Lewis, S. L., & Maslin, M. A. (2018). *The human planet: How we created the anthropocene* (International Edition). Yale University Press.

Lower, E., & Sturtevant, R. (2021). *Alien language: The importance of metaphor*. Michigan Sea Grant Extension Office. Retrieved October 5, 2021, from https://www.canr.msu.edu/news/alien-language-the-importance-of-metaphor-msg21-sturtevant21

Marbuah, G., Gren, I. M., & McKie, B. (2014). Economics of harmful invasive species: A review. *Diversity, 6*(3), 500–523.

Martin, C. (2013). Turbulent stillness: The politics of uncertainty and the undocumented migrant. In D. Bissell, G. & Fuller (Eds.), *Stillness in a Mobile World* (pp. 192–208). Routledge.

Martin, C. (2016). *Shipping container*. Bloomsbury Publishing.

Mee, K., & Wright, S. (2009). Geographies of belonging. *Environment and Planning A, 41*(4), 772–779.

Miller, A. W., & Ruiz, G. M. (2014). Arctic shipping and marine invaders. *Nature Climate Change, 4*(6), 413–416.

Molnar, J. L., Gamboa, R. L., Revenga, C., & Spalding, M. D. (2008). Assessing the global threat of invasive species to marine biodiversity. *Frontiers in Ecology and the Environment, 6*(9), 485–492.

Ojaveer, H., Galil, B. S., Carlton, J. T., Alleway, H., Goulletquer, P., Lehtiniemi, M., Marchini, A., Miller, W., Occhipinti-Ambrogi, A., Peharda,

M., Ruiz, G. M., Williams, S. L., & Zaiko, A. (2018). Historical baselines in marine bioinvasions: Implications for policy and management. *PLoS One, 13*(8), e0202383. https://doi.org/10.1371/journal.pone.0202383

Peters, K. (2010). Future promises for contemporary social and cultural geographies of the sea. *Geography Compass, 4*(9), 1260–1272.

Pimentel, D. (2002). Non-native invasive species of arthopods and plant pathogens in the British Isles. In D. Pimentel (Ed.), *Biological invasions: Economic and environmental costs of alien plant, animal, and microbe species* (pp. 151–158). CRC Press.

Psuty, N. P., Steinberg, P. E., & Wright, D. J. (2004). Coastal and marine geography. In G. L. Gaile & C. J. Willmott (Eds.), *Geography in America at the dawn of the 21st century* (pp. 314–325). Oxford University Press.

Ricciardi, A., & MacIsaac, H. J. (2011). Impacts of biological invasions on freshwater ecosystems. In D. M. Richardson (Ed.), *Fifty years of invasion ecology: The legacy of Charles Elton* (pp. 211–224). Blackwell.

Robbins, P. (2004). Comparing invasive networks: Cultural and political biographies of invasive species. *Geographical Review, 94*(2), 139–156.

Robbins, P., & Moore, S. A. (2012). Ecological anxiety disorder: Diagnosing the politics of the anthropocene. *Cultural Geographies, 20*(1), 3–19.

Ruiz, G. M., & Carlton, J. T. (2003). Invasion vectors: A conceptual framework for management. In G. M. Ruiz, & J. T. Carlton (Eds.), *Invasive species: Vectors and management strategies* (pp. 459–504). Island Press.

Saebi, M., Xu, J., Curasi, S. R., Grey, E. K., Chawla, N. V., & Lodge, D. M. (2020). Network analysis of ballast-mediated species transfer reveals important introduction and dispersal patterns in the Arctic. *Scientific Reports, 10*(1), 19558. https://doi.org/10.1038/s41598-020-76602-4

Savitzky, S. (2016). *Icy futures: Carving the Northern Sea Route* (Doctoral Dissertation). Lancaster University.

Soto, I., Balzani, P., Carneiro, L., Cuthbert, R. N., Macêdo, R., Serhan Tarkan, A., ... & Haubrock, P. J. (2024). Taming the terminological tempest in invasion science. *Biological Reviews, 99*, 1357–1390.

Secretariat of the Convention on Biological Diversity. (2020). *Global biodiversity outlook 5*. Montreal.

Seebens, H., Gastner, M., & Blasius, B. (2013). The risk of marine bio invasion caused by global shipping. *Ecology Letters, 16*, 782–790.

Smout, T. C. (2014). What's natural: A species history of Scotland in the last 10,000 years. *Glasgow Naturalist, 26*(Part 1), 11–16.

Steinberg, P. E. (1999). Navigating to multiple horizons: Towards a geography of ocean space. *Professional Geographer, 51*(3), 366–375.

Steinberg, P. E. (2001). *The social construction of the ocean.* Cambridge University Press.
Steinberg, P. E., & Peters, K. (2015). Wet ontologies, fluid spaces: Giving depth to volume through oceanic thinking. *Environment and Planning D: Society and Space, 33*(2), 247–264.
Streftaris, N., Zenetos, A., & Papathanassiou, E. (2005). Globalisation in marine ecosystems: The story of non-indigenous marine species across European seas. In R. N. Gibson, R. J. A. Atkinson, & J. D. M. Gordon (Eds.), *Oceanography and marine biology: An annual review* (pp. 429–464). CRC Press.
Sweet, L. C., Green, T., Heintz, J. G. C., Frakes, N., Graver, N., Rangitsch, J. S., Rodgers, J. E., Heacox, S., & Barrows, C. W. (2019). Congruence between future distribution models and empirical data for an iconic species at Joshua Tree National Park. *Ecosphere, 10*(6), e02763. https://doi.org/10.1002/ecs2.2763
Tsing, A. L. (1995). Empowering nature, or: Some gleanings in bee culture. In S. Yanagisako & C. Delaney (Eds.), *Naturalizing power: Essays in feminist cultural analysis* (pp. 113–143). Routledge.
Urry, J. (2007). *Mobilities* (1st Edition). Polity.
Warren, C. R. (2021). Beyond 'native v. alien': Critiques of the native/alien paradigm in the Anthropocene, and their implications. *Ethics, Policy & Environment, 26*(2), 287–317.
Zenni, R. D., Essl, F., García-Berthou, E., & McDermott, S. M. (2021). The economic costs of biological invasions around the world. *NeoBiota, 67*, 1–9.

Human-Shark Encounters Beyond Borders: (Post-humanist) Attempts to Navigate a Maritime Contact Zone

Julia Verne

Introduction: How Shark Attacks Challenge Ocean Governance

Apparently, you are 15 times more likely to be killed by a coconut falling from a tree while lying comfortably on the beach than being attacked by a shark. At least, this is what Burgess, the director of the Florida Museum of Natural History's International Shark Attack File stated in 2002, referring to a press release of a British travel insurance company in which they guaranteed holidaymakers full cover if they were hit by falling coconuts (Selingo, 2002). Although it has since been revealed that the notion that coconuts kill around 150 people worldwide each year was falsely derived from a four-year review of trauma admissions to a Provincial Hospital in Papua New Guinea published in 1984 (Barss, 1984), for which the author later received an Ig Nobel Award by the *Annals of Improbable Research*, this legendary statistic still holds strong

J. Verne (✉)
Johannes Gutenberg University Mainz, Mainz, Germany
e-mail: julia.verne@uni-mainz.de

K. Peters and J. Turner (eds.), *Ocean Governance (Beyond) Borders*,
https://doi.org/10.1007/978-3-031-71322-4_9

in the public opinion. And, independent of the dubious assumptions behind this global statistic, compared to other potential dangers encountered in aquatic recreation, shark attacks still appear to be the least likely according to the regularly updated database of the International Shark Attack File hosted by the Florida Natural History Museum.

In line with this, the Professional Association for Diving Instructors (PADI) recently announced the 16 best sites to dive *with* sharks (Shofield, 2024). Others, who are more concerned about their safety, may opt for numerous offers of shark cage diving in which metal cases are used to maintain a clear separation between humans and sharks. Being in close vicinity to these potentially deadly animals, which have long featured prominently as 'killing machines' in popular culture, is supposed to replace fear by fascination and encourage an understanding of the need to protect increasingly endangered sharks.

In 1991, a Shark Specialist Group was initiated as one of the International Union of the Conservation of Nature's (IUCN) Species Survival Commissions which continually assesses the IUCN Red List status of all species of sharks, rays and chimaeras and develops conservation strategies. In relation to this, a number of international organisations have since made an effort to foster the protection of sharks through science and education. As the WWF states 'sharks are in crisis', with more than a third of all species threatened with extinction mainly due to overfishing. Bringing people closer to sharks is generally regarded as an important step to gain support for their protection (WWF Mediterranean Marine Initiative, 2019).

However, on the island of La Réunion, the proximity between humans and sharks has led to another kind of 'shark crisis' (locally called 'la crise requin') (Thiann-Bo Morel, 2019). In this French Overseas Department in the Indian Ocean, a total of 27 unprovoked shark attacks have been recorded since 2011, eleven of which were fatal to humans (Habiter la Réunion 2020, Centre Sécurite Requin 2024). The majority of the victims were surfers, while others were in the coastal waters with bodyboards, snorkels or in a dugout canoe. Here, all of a sudden, it seemed much more likely one could be attacked by a shark rather than being hit by a falling coconut—an idea that people in La Réunion were not ready to accept. This situation led to a highly polarised controversy about different ways to govern the coastal waters of the island. Apart from some who called for a radical shark culling program, hoping that this would significantly reduce the number of sharks in this part of the ocean, the majority of the measures discussed centred around attempts to separate humans and sharks. However, it soon became clear that—in large part due to the specific materiality of the ocean which brings with it particular logistical demands and makes such bordering attempts even more costly—it would be impossible to assign separate spaces for sharks and humans demarcated and maintained by a clear border and that some compromises had to be made. This situation in La Réunion resonates with contemporary conceptual debates that highlight the limits of a clear human-nature divide which has dominated much of scientific thought since enlightenment.

Post-humanist and 'more-than-human' approaches in particular emphasise the symbiosis of lifeworlds and the resulting co-production of landscapes (and seascapes), showing how the human and non-human cannot be clearly distinguished from each other either empirically or conceptually (see, among others, Barad, 2003; Haraway, 2008; Whatmore, 2014). In contrast to a dualistic view of spaces as either 'natural' or 'cultural', human or non-human, Haraway famously used the concept of '"contact zones", ... "mortal world-making entanglements" ... in which those who are to be in the world are constituted in intra- and interaction' (Haraway, 2008: 4). In this contribution, I examine the 'shark crisis' in the coastal waters of La Réunion as a contact zone in which human orderings of space are challenged and negotiated.[1] After a brief overview of events, I particularly emphasise the different attempts to govern this ocean space through various bordering technologies. While some technologies are intended to maintain a clear separation between sharks and humans (e.g. shark nets), others, such as tracking technologies, follow a more relational approach, however, still with the aim to

[1] This text is developed out of my own empirical observations during a two-month stay on the island of La Réunion as a visiting scholar at the Université de la Réunion in 2013 as well as subsequent analysis (social) media discussions on the island's 'shark crisis' between 2011 and 2020 and an engagement with the different measures developed and introduced by the Centre Sécurité Requin in light of current conceptual debates on human-technology-nature relations.

avoid encounters. In contrast, as I will illustrate, several more recent technological innovations resonate with post-humanist approaches as they try to enhance or alter human capabilities so that they do not appear as a potential bait anymore and, thus, can share the maritime space without fear. Overall, the different shark-bite mitigation strategies developed and employed in La Réunion provide a vivid insight into how such diverse technologies mediate and reorganise the relationship between humans and sharks, both discursively and practically. They also indicate the persistent reliance on 'technological fixes' in ocean governance (Nyman, 2019; Weinberg, 1993). However, by highlighting the specific challenges of creating and maintaining fixed boundaries in the ocean, they encourage us to envision ocean governance beyond borders while still striving for a less risky co-existence for both humans and sharks which is considered vital for the health of the ocean.

The Shark Crisis in La Réunion: A Brief History of Events

The presence of sharks in the coastal waters of La Réunion has always been well known to the local population, commonly referred to as Créole. Nevertheless, when surfing and the associated surf culture became popular in the 1960s, French mainlanders introduced the sport to the Indian Ocean island and were soon joined by islanders. La Réunion became well known for ideal surfing conditions, with Surfing Magazines and films advertising the best spots in the 1990s. This was in spite of 20 shark attacks—ten of which were fatal—being recorded throughout this decade. However, with only one fatal attack between 2000 and 2010, it came as a shock to the island when six unprovoked shark attacks occurred in 2011 alone (Lemahieu et al., 2017). This year marked the start of what became called 'la crise requin' ('the shark crisis') of La Réunion. The victims included both local and overseas surfers and bodyboarders. Two years later, in which eight more attacks were observed, the government decided to ban swimming and surfing on most of the island. Swimming and water sports such as snorkelling or Stand

Up Paddling only remained allowed in the lagoons. Entering the water outside of these lagoons became illegal and subject to a monetary penalty.

While the majority of people obeyed the new law, many of them did not agree with this measure. Under the slogan 'Rend a nou la mer, rend a nou la Réunion' ('Give us the sea back, give us Reunion back'), a number of protests were initiated as people felt that their right to the sea had been taken away (Sorgue, 2017). As the initiators of this movement put it: 'How can you imagine an island without the ocean?'. With the 'shark crisis' leading to a steep decline in tourism and related economic sectors, the government's decision to solve the issue by simply separating humans and sharks had severe implications for a large part of the population. Moreover, when in 2015 a 13-year-old local boy and promising surfer could not resist his passion and was killed by a shark when surfing despite the ban, public protests reached their peak. They demanded an enhanced effort by the government to come up with and implement some more sustainable measures that would also account for the economic and ecological effects of the shark crisis.

Overall, despite the ban, 26 unprovoked shark attacks were recorded in La Réunion between 2010 and 2020, ten of them fatal. In the period from 2011 to 2016, the island accounted for 16% of shark attacks worldwide and, with 3.15 shark-related deaths per 1 million people, it had by far the highest rate in the world (Chapman & McPhee, 2016; Lagabrielle et al., 2018). The shark attacks were mainly caused by both bull sharks (*Carcharhinus leucas*) and tiger sharks (*Galeocerdo cuvier*), which have been banned from fishing since 2007, when the west coast of the island was declared a marine protected area. Some see this nature reserve and the associated fishing ban as the main cause of this 'aquatic emergency' (Guinand & Henchoz, 2016), leading to demands to revoke the conservation status. However, increased urbanisation and the associated pollution of coastal waters through runoff, overfishing and a resulting lack of sufficient food for the sharks, as well as a decline in the more sensitive reef shark population and the takeover of their territories by bull and tiger sharks, are also considered possible reasons for the apparently 'extremely high density of aggressive large sharks' off the coast of La Réunion (Fédération Française Surf, 2015: n.p.), which is situated along the so-called 'shark highway' between South Africa and Australia.

In order to find out more about the reasons for the increased number of incidents and to assist political decision-making, the government soon founded a committee of politicians, scientists and members of the public (Comité Réunionnais de reduction du risqué requin, Co4R) which was meant to address the issue and also initiated a research project focusing on biological and ecological questions concerning the habitat use of the two shark species (CHARC 2011–2016) (Lemahieu et al., 2017). However, in light of continuing shark attacks, this programme was highly contested. To bring together different perspectives, a more comprehensive resource centre was funded by the government in 2016, which subsequently formed the Centre Sécurité Requin (CSR) in 2020, where politicians, community groups and scientists continue to work on finding appropriate measures to solve the island's shark crisis. Over the years, they have employed a number of different technologies (discussed later), some taken and adapted from other places, some developed locally, which have led to a complex system of shark-bite mitigation strategies allowing humans a (highly regulated) return to the sea. With the last shark attack recorded in 2019, the number of swimmers and surfers has increased and even a couple of surf shops have been opened again. Nevertheless, a tension remains on the island, including heated debates and complaints about the way in which the government has been handling the situation.

Governing Human-Shark Encounters Through Bordering Technologies

Human-wildlife conflicts have long been the focus of various research perspectives, which are often based on a dualistic opposition of nature and culture. Approaches from the field of conservation biology and wildlife management are often centred on the assumption that global population growth and the associated encroachment into previously uninhabited landscapes has driven up the number of conflicts (Woodroffe et al., 2005). The analyses usually focus on the quality of the damage (human victims, livestock damage, damage to crops), various risk management techniques (lethal control, protected areas) as well as recommendations for action and related policies (global forums for nature conservation, transregional cooperation, etc.) (Knight, 2000). While wildlife conservation is primarily measured in terms of its ecological benefits and the need to preserve wildlife habitats (Brooks et al. 2006), when it comes to human-wildlife conflicts it is usually the use and security requirements of humans which are placed centre-stage (Löe & Röskraft, 2004). Depending on whether the focus is on the 'cultural space' threatened by wild animals or the 'natural space' worthy of protection, the measures usually concentrate on excluding either the one or the other from the space, thus achieving a separation between humans and nature (Brockington & Igoe, 2006; Gardner, 2016).

Often enough, however, it becomes apparent that the distinction between spaces made by humans is dissolved by the mobility of animals. Despite numerous efforts to achieve co-existence through separation, contact zones are ubiquitous. On the one hand, in many places, wild animals resist enclosure and containment: they are mobile, know no national borders and move within and outside the territories ascribed to them (see e.g. Barua, 2014; Lorimer, 2000; Whatmore & Thorne, 2000). As Maxwell et al. (2015) have shown, this is particularly pertinent to discussions of the sea, where species are mobile in a fluid environment. On the other hand, the experience of wilderness is also playing an increasingly important role in leisure activities. This also characterises the situation in the coastal waters of La Réunion where the rising number of human-wildlife encounters due to the popularity of water sports is challenging the separation of 'nature' and 'culture'.

Shark Culling

Conventional measures to ensure a separation of humans and wildlife are often based on the assumption that co-existence in this disputed territory is not possible. If, for example, a shark is sighted or someone is killed by a shark, debates explode in the (social) media about whose territory this is, who has the right to be there and who is responsible for enforcing this right (see e.g. Gibbs, 2018). In such situations, there are still frequent attempts at a so-called lethal control of wild animal populations and, in

La Réunion too, demands for shark culling featured prominently among the protesters.

Since 2014, the government of La Réunion has indeed conducted four different shark fishing programmes, following the examples of Australia and South Africa, where shark culling has been practised regularly for more than 70 years (Gibbs & Warren, 2015). Despite the popularity of such programmes among parts of the public, there is a strong opposition of environmentalists and conservationists who point out the ecological costs and ethical issues of killing sharks (see. e.g. Neff, 2015). Moreover, even though some may have hoped that regular shark fishing would be enough to claim back the ocean territory that had been left to the sharks with the implementation of the marine protected area (FFS 2018), as Williamson (2015) highlighted 'there is no scientific support for the concept that culling sharks in a particular area will lead to a decrease in shark attacks and increase ocean safety' (see also Holland et al., 1999 and Wetherbee et al., 1994 regarding the effects of shark culling in Hawaii between 1959 and 1976). Thus, while the fishing of potentially dangerous sharks continues, the Centre Sécurité Requin also invested in the (further) development of other measures to keep humans and sharks apart.

Shark Nets

Fences are still one of the most widely used technologies to enable the co-existence of humans and wild animals by means of spatial separation. On land, ordinary fences made of wire, rope or wood are commonly used to maintain clear boundaries, sometimes enhanced with the help of electrical voltage or resistant materials. More recently, virtual fences, which function with transmitter collars but without physical barriers, represent new developments especially for the containment and demarcation of grazing animals (Jachowski et al., 2014). Similarly, fencing technologies have been developed for use in the ocean. Shark nets were first employed in New South Wales, Australia, in 1937 to exclude sharks from swimming areas, and their specific shape and design have been continuously developed since then. However, contrary to the widespread perception that the nets create a completely enclosed area, most programmes actually involve fishing nets that are several hundred metres long and a couple of metres wide and are set in water depths of 10–12 metres at a distance of 500 metres from the shore in order to catch sharks (Green et al., 2009). As Gibbs emphasises:

> Importantly, the nets do not create a barrier between swimmers and the open ocean. There is no 'inside' or 'outside' a netted zone keeping people and sharks apart. ... The nets do not form territory by creating bounded space. Rather, they function through social power. The *Shark Meshing Program* produces territory by creating a sense of human control and authority over the entire coastline, and enacting violence against nonhuman animals. (Gibbs, 2018: 208)

Nevertheless, in Boucan Canot and Roches Noires, two of the most popular beaches in La Réunion, a unique net system was developed in 2016 that can be folded manually during storms (Guinand & Henchoz, 2016: 8). The nets were meant to keep an area of 140,000 square metres 'guaranteed shark-free'. However, only a few months after its installation, there was a hole in the net and a shark attack took place in one of the secured areas (Surmont, 2016). Since then, two more holes—attributed to sabotage—have been discovered in the nets, indicating how such borderings cannot just be implemented once but need frequent maintenance and repair as they are exposed to both social and physical forces that challenge their stability.

In two test phases in 2018 and 2020, the CSR also installed the so-called SharkSafe Barrier technology (SBB). Promoted as 'the first nature-inspired eco-friendly technology to protect people and sharks' by its designers (Sharksafe Barrier, 2021: n.p.), the barrier uses biomimicry by creating the visual effect of a kelp forest while also generating a strong magnetic field through ceramic magnets. However, even though no shark was observed to pass through the barrier, in one spot most of the tubes anchored to the seabed were not able to withstand the swells on the islands. As the Head of one of the shark fishing projects in La Réunion admitted: 'The sea band along the coast is not a closed swimming pool'

(Guinand & Henchoz, 2016: 8). Thus, even though several communities on the island still have shark nets in place, they also state that 'the presence of the net alone does not mean that the safety system is operational' (CSR, n.d.: n.p.). Every morning an inspection of the conditions takes place before the decision is made if the perimeter is actually opened for bathing.

In addition to stationary fences, there are also other technologies to keep sharks and humans apart. This is done, for example, with the help of so-called drum lines, ropes equipped with bait on large hooks. A smart drum line was developed in La Réunion, which has been in use since 2014 to detect the sharks at the baits more quickly and thus increase the chances for their survival. Solar-powered GPS transmitters have been installed on the hooks, enabling the animals caught to be reached alive in 86% of cases. The sharks that are released are tagged, and those that are killed are scientifically examined.

However, many have feared that these lines do not increase the safety of swimmers and surfers by keeping the sharks away from the coast, but instead attract them with the bait (Guinand & Henchoz, 2016: 8). However, based on movement data in different parts of the world where such technologies have been used, no significant difference can be identified between the time before the lines were installed and today (Gibbs, 2018; Wetherbee et al., 1994).

Shark Surveillance

In addition to such smart drum lines which alert the presence of sharks, a baited remote underwater video system as well as a sonar system and drones are tested to surveil the coastal waters for sharks. Whereas, for example, in Cape Town, Shark Spotters are positioned at an elevated level at the coast to look out for approaching sharks, in La Réunion the setting has proved to be difficult for detecting sharks on time. As the CSR admits, 'numerous aerial images taken by paragliders, microlight pilots and airplane pilots over Reunion's coastline have already revealed the presence of large predators, albeit in a delayed manner' (CSR, n.d.: n.p.). The technical requirements for a drone to successfully spot sharks in the water are very high. Yet, despite constraints linked to the nature of the seabed, lighting conditions, wind and turbidity, they are still used as a complementary tool to surveil the waters from above. Finally, in association with an international engineering company, the CSR is currently working on the development of an autonomous camera equipped with a shark detection algorithm with the aim to deploy this detection camera system as a surveillance tool in selected surf sports spots in addition to patrol boats.

The spatial bordering logic of these different technologies is also expressed in the 'zones of experimentation (ZONEX)' (CSR, n.d., see also Gibbs, 2018 for 'making territory' through the shark meshing programme in New South Wales, Australia) which have been announced by the government in which human access to the sea should become possible again through the combination of different shark safety measures. Nevertheless, even though at first glance, fences and trap lines in maritime areas may promise increased control and security, when scrutinised it becomes clear that they often fail to deliver. They attempt to separate spaces and thus avoid encounters, but, as the tests of the different technologies mentioned above show, they are much more likely to still form a contact zone in which the relationship between humans and sharks is existentially negotiated. In effect, dealing with co-existence in the coastal waters of La Réunion seems unavoidable, but on what terms?

Post-humanist Approaches to the Ocean

The ocean is one of the last spaces on earth about which humans have only very limited knowledge and into whose depths they have not yet fully penetrated (Rozwadowski, 2018). Due to the special materiality of the ocean and its associated inaccessibility to humans—at least without oxygen—the ocean has long been considered a natural space (Steinberg, 1999), or, as Mack put it, 'a quintessential wilderness' (Mack, 2011: 17). As Squire's (2021) vivid engagement with the undersea living projects of the US Navy illustrates so well, it is only with the help of various underwater technologies is it even possible to expand our knowledge of the ocean and underwater life. Well-equipped divers, deep-sea expeditions with submarines and underwater robots are now exploring selected areas underwater, taking samples and collecting data (Laloë, 2014, 2016; Picken & Ferguson, 2014). However, these too can only provide very limited insights due to the depth, the size of the area and the constant movement of the water masses. The focus of these efforts is therefore often the targeted search for valuable raw materials, the monitoring of coral reefs and other maritime conservation areas, or the search for shipwrecks (Bremner, 2014; Helmreich, 2007; Lehman, 2018).

In effect, as Steinberg has pointed out in one of the early writings introducing a human geographic approach to the sea, especially with regard to its resources the ocean has been presented as a 'frontier replete with opportunity, at last capable of being "conquered"' (Steinberg, 1999: 404). And indeed, not only within attempts to increase our knowledge about the ocean and its maritime resources but also as part of the increasing popularity of different kinds of water sports, in the last decades, humans have ventured further and in larger numbers into the sea (see Ingersoll, 2016 for an engagement with the effects of this 'colonization of the sea' and neocolonial Surf Tourism in Hawai'i). As Gibbs (2018: 213) has highlighted in her engagement with shark-bite mitigation policies in New South Wales, Australia, through 'recreational and professional activities humans use the water, seeking to claim a territory of sorts', 'to establish a place there' (Gibbs, 2018: 217; see also Anderson, 2012; Waitt & Warren, 2008). As Gibbs notes,

> As individuals and as a society we seek power and authority over the coast through our material interactions; we claim territory with our bodies, boards, wetsuits, fins, masks, breathing apparatus, lifeguards, observation towers, jet-skis, beach buggies and boats. (Gibbs, 2018: 214)

On the other hand, conservationists in particular emphasise the sea as a non-human world in which humans are mostly a harmful intruder. From this point of view, 'in entering the water [humans] are entering sharks' territory' (Gibbs, 2018: 209).

In this struggle over territorial control, the shore as a liminal space between land and sea forms a 'complex system of diverse intersecting mobilities' (Sheller & Urry, 2004: 6). As Wilson has convincingly shown in her analysis of the BBC Blue Planet II series, the sea has become a space 'where different forms of knowledge, technology, people, and non-human life grapple with each other in conditions that are shaped by shifting forms of power' (Wilson, 2019: 712). This also becomes apparent when looking at the situation in La Réunion where encounters between water sports practitioners and sharks become more and more regulated within attempts to successfully manage and solve the 'shark crisis' by recurring to different forms of knowledge and technologies.

To examine spaces of intervention—the 'space and time where subjects previously separated by geography and history are co-present' (Pratt, 1992: 6)—Pratt (1991) introduced the concept of the contact zone. While her original aim was to shift the attention from studies of imperial centres to actual spaces 'where cultures meet, clash and grapple with each other, often in contexts of highly asymmetrical relations of power such as colonialism, slavery, or their aftermaths' (Pratt, 1991: 34), the contact zone has recently also been used to think about encounters with the more-than-human (see e.g. Barua, 2014; Sundberg, 2006). Most famously, the contact zone has been taken up by Haraway (2008) in her post-humanist analysis of the agility training she did with her dog in which she foregrounds the deeply entangled relations between 'companion species'. In a recent themed issue, Isaacs and Otruba (2019) suggest taking the concept of the contact zone to environmental research more generally in order to be more attentive towards more-than-human agency, to address issues of violence and injustice and contribute to the decolonisation of knowledge production. As Wilson points out in the same issue, 'the use of the contact zone as an analytical lens should draw attention to complexity: different configurations and forms of power as they are reworked to different effects' (Wilson, 2019: 726). In this regard, the contact zone does not prioritise the encounter, 'but rather concerns multiple forms of relation, communicative practice, and contact, and explicitly focuses on spaces that are shaped by significant asymmetry' (Wilson, 2019: 718).

Indeed, the encounters between humans and sharks in the coastal waters of La Réunion are shaped by clear asymmetries when it comes to the ways in which both navigate the maritime space. On the one hand, this concerns aspects such as movement and sight in the water itself, on the other hand, it also entails their unequal representation in the political arena which also effects the conditions of the contact zone. As Carter and Palmer have argued, 'the human construction and enforcement of … boundaries reflect an asymmetry of power founded on the principle of human exceptionalism' (2016: 4). Through the rising number of attacks and their observed presence, the sharks, however, appear to resist any marginalisation and make it difficult to ignore them 'in defense of a stable, centered sense of knowledge and reality' (Pratt, 1991: 37). To many, the threat of 'shark attacks' threatens assumed boundaries between a safe beach and the wild open waters, leading, as I have sketched out above, to radical attempts to avoid encounter and maintain a categorial separation (Taglioni & Guiltat, 2015; Todd & Hynes, 2017). Nevertheless, the ongoing transgressions (from both sides) of the diverse forms of boundaries drawn through different measures show the impossibility of complete control. As Wilson (2017) has pointed out, it is exactly the rupturing of such borders and the resulting encounters in the contact zone that may 'open up a site of ethical, pedagogical, and political potential' (Wilson, 2017: 718). Engaging with Dingoes on Fraser Island, Carter and Palmer have illustrated how human-animal encounters may shape new ethical topologies and 'a willingness by humans to disrupt spatial and cultural boundaries and the boundaries between thinking and feeling, in order to shape more responsive, respectful and less anthropocentric topologies' (2017: 13).

Such post-humanist decentring of humans plays a decisive role in contact zones as envisioned by Haraway (2008). Building on Pratt's relational understanding of contact zones as they 'emphasize how subjects get constituted in and by their relations to each other' (1992: 8), Haraway focuses even more on 'the taste of copresence and the shared building of other worlds' (2008: 237). As Haraway claims, many of the 'transformative things in life happen in contact zones' as they change all subjects involved 'in surprising ways' (2008: 219). Yet, while it may be fairly easy for many to agree to the 'shared building of other worlds' with regard to rather common companion species such as dogs, or dolphins (Cloke & Perkins, 2005; Taylor & Carter, 2013), the co-existence with animals perceived as 'incompanionate' (Carter & Palmer, 2017) appears to be even more demanding (notwithstanding that dog bite fatalities, according to statistics compiled by the Florida Museum of Natural History, amounted to 349 compared to 65 fatalities due to sharks in 2021). As Haraway puts it, 'if we appreciate the foolishness of human exceptionalism, then we know that becoming is always becoming with—in a contact zone where the outcome, where who is in the world, is at stake' (2008: 244). And indeed, as the following ways of dealing with the 'shark crisis' in La Réunion show, a less anthropocentric approach to the coastal waters includes a willingness to take this risk.

Becoming More-Than Human?

Beyond classic, even if sometimes recently adjusted technologies of separation and efforts to achieve more control over the coastal waters through different surveillance technologies, in recent years, there have been increasing attempts to develop technologies that make it possible to go beyond clear human-non-human distinctions. On the one hand, this includes measures that include a humanisation of sharks. In Western Australia, for example, more than 400 sharks have been tagged after they have been caught with the help of smart drumlines. Via satellite, the mobility of these sharks becomes accessible. Under the motto 'Switch on your sea sense', the related Shark Smart app offers insights into this mobility (www.sharksmart.com.au). This is supplemented by information on those sections of the beach which are patrolled by Surf Life Saving and a weather forecast, which also includes warnings of strong currents and particularly strong swells. Some of the tagged sharks are even given a name and their own Twitter (now X) account to regularly 'Tweet' their location and some short messages. This way, beachgoers and others interested in sharks are able to learn more about sharks from the sharks themselves as they are able to follow them virtually. In turn, the sharks notify beachgoers of their presence by automatically sending alerts as soon as they reach a certain proximity to the beach. While such animal-generated geodata form an important step towards what has been called a 'transparent nature' (Benson, 2010), the vision of a seamless monitoring of the non-human environment is still a long way off and entering the water remains risky.

Nevertheless, even when surfing and swimming was officially banned on La Réunion, there was a group of five to seven surfers who kept going surfing every day despite their fear of sharks (see also Gegan, 2023). As Laury Le Costour famously stated in an interview: 'We've lost some friends around us. It was a nightmare, but the passion goes on and we still go. We were still aware of the risk. At that time it was *Roulette Rousse*, Russian Roulette. It can happen to you' (Le Costour in Gegan, 2023: n.p.).

Sceptical of the effects of the bordering technologies in place, they preferred to rely on human eyes. With friends on the look-out, they entered the water at times thought to be less risky—i.e. avoiding dirty waters or times of dawn when sharks were observed to come closer to the beach and visibility would be even more limited. On the one hand, this indicates how contact zones are not always stable but may be characterised by certain patterns and rhythms (Anderson, 2012). On the other hand, this places emphasis on the need to be able to 'read' the sea (Ingersoll, 2016: 34).

In order to further increase the safety of surfers, some divers and surfers have since been trained as so-called *vigies*. With the help of fins and snorkels, these underwater shark spotters move in the water like fish but are equipped with harpoons. If they see a shark, they are supposed to pass this information on to one of the patrol boats. The idea for this came from a surfer who wanted to protect his nephew by observing him underwater while he was taking part in a surfing competition in 2011. Since then, a formalised training for *vigies* has been developed on the island making them part of the official measures (CSR, n.d.). Meanwhile, *vigies* have become a central part of the complex assemblage of different technologies that support the co-existence of humans and sharks. In contrast to the other measures sketched out above, as Basson points out, for the *vigies* this indeed entails a renegotiation of the relation between humans and sharks as they give up human dominance and enter the water on more equal ground. As one of the *vigie*s recalls:

> Thanks to the surf resistance on our island and to those who were ready to die for our passion, we were able to develop unique structures that today allow people to go back and enjoy the water once again with an acceptable risk. ... Vigie, it's not only our job but our passion. When our eyes are under the surface, we know our role and there is not an ounce of fear in us. We couldn't be further from a prey, and our sole presence in the water column is enough to deter the predators to approach the zone. (Basson cited in Guinand & Henchoz, 2016: 7)

Furthermore, in order to prevent sharks from perceiving the human body as potential prey, various companies have developed innovative wetsuits in recent years that either imitate the reflection of light on water and thus make people look 'water-like', or adapt the appearance of people to the appearance of poisonous fish (see Radiator Waterwear, 2019). Although initial tests have been promising, there is still no clear evidence that this way of going beyond the human appearance actually protects against a shark attack. There is also a focus on researching different repellents and technologies that people can carry with them to keep sharks at bay. For example, some companies, like NoShark, RPELA, SHARKBANZ and Ocean Guardian, sell small battery-powered devices—sometimes in the form of a wristband or to be worn around the ankle—that constantly emit a high voltage signal in wave form, which sharks find unpleasant due to their highly sensitive Lorenzinian ampullae (Blount et al., 2021; Gibbs & Warren, 2012). Other technologies, such as Podi, use scents that mimic the smell of a dead shark and are released evenly into the water to encourage a shark to swim away. This is based on traditional methods used by fishermen in the Indian Ocean, who use a dead shark in tow to protect their catch from other sharks. Instead of attempting to create fixed borders in a fluid environment, this indicates the potential of less material but more dynamic, sensory borders at sea.

In the past years, in addition to the other measures and the *vigies*, all surfers entering the water at St. Leu, a famous surf spot at the west coast of the island, have to register at the beach and show that they have a shark deterrent device on them. Thus, in a certain way, the surfers become more-than-human when entering the water, changing their visual appearance through particular wetsuits, their electromagnetic fields or scent. Together with their surfboards or fins, it becomes possible for them to delve into the water again and experience a particular sense of convergence with the sea (Anderson, 2009, 2012, 2014).

Maritime Contact Zones Beyond Borders?

> There are no simple government solutions when sharks bite people. These rare and sometimes fatal incidents are fraught with uncertainties and command a disproportionate amount of psychological space in the minds of the public, as well as a large degree of policy space and funding from many governments. (Neff, 2012: 88)

The situation in La Réunion over the past decade shows exactly this: fierce debates about appropriate measures to secure public safety (Losen, 2023). Together with the urgency to counteract the economic decline of the island due to the extreme decrease in tourism as a result of 'the shark crisis', these have led to the development and implementation of a variety of different technologies aiming to mitigate the risk of shark bites. As this chapter illustrates, these different technologies reveal interesting aspects of current ocean governance and the relation between humans and the sea more generally. They show the need to acknowledge the ocean 'as a *more-than-human* space; taking its very nature into account when examining social processes which occur on and under its surface' (Peters, 2014: 177).

Whereas the government was first hoping to be able to solve the issue with a simple ban, keeping people out of the ocean, it soon became clear that this kind of separation would not be accepted by a large part of the local population (see e.g. Thiann-Bo Morel, 2019). Instead, people lay claim to the coastal waters and did not want to leave them to the predators. As Pinault has emphasised: 'The ocean is the shark's territory, and we don't belong there. It's the first logical judgement. But here on Réunion Island, it's a little more complicated than that' (Pinault cited in Guinand & Henchoz, 2016: 7).

On the island of La Réunion, in a series of protests, numerous people made clear that they do not want to share its coastal waters with dangerous wild animals. However, against the backdrop of debates about species protection, animal ethics and biodiversity, it seems impossible to completely ignore animal welfare in favour of human welfare. As a result, the coastal waters of La Réunion form a contact zone 'where different forms of knowledge, technology, people, and non-human life grapple with each other in conditions that are shaped by shifting forms of power' (Wilson, 2019: 712). In order to minimise potentially deadly encounters, different technologies are currently being used to creatively and experimentally deal with the boundaries between 'cultural' and 'natural' space, humans and wildlife. Highlighting the specific challenges of creating and maintaining fixed boundaries in the ocean, they encourage us to envision ocean governance beyond borders while still striving for a less risky co-existence for both humans and sharks, which is considered vital for the health of the ocean.

The effects of the various technological innovations that are (to be) used are often not yet clear—but they are certainly far from proving a simple technological fix to the issue. La Réunion can currently be seen as an experimental laboratory in search of technical solutions to a problem that is existential for both sides. With statistics still placing La Réunion as one of the 'sharkiest' spots in the world, it is considered the ideal test site for different shark deterrents. A focus on these technologies allows for a better understanding of the complex networks of relationships that are currently emerging in these contact zones which appear to be characteristic of a renegotiation of the relationship between humans and nature in the Anthropocene more generally.

Whereas the common technologies of separation aim to consolidate the power position of humans in divided spaces, the use and development of various more-than-human mediators point in the direction of a 'politics of conviviality' (Hinchliffe & Whatmore, 2009: 106). The *vigies*, the change in physical appearance through certain wetsuits or the imitation of the smell of death show particularly clearly how the post-humanist concern of convivial co-existence may alter the relationship between humans and sharks, both discursively and practically. However, the extent to which humans actually relinquish their position of dominance and instead place themselves on a level with wild animals remains very different. This is not just about changing one's appearance, but conceptually also involves a certain biologisation of the human, as well as the contact zones themselves in which humans and sharks are supposed to live 'alongside' each other (Latimer, 2013: 77). Technologies here are therefore not independent, separate, mere technical devices that are placed between humans and wild animals with the intention of preventing encounters in the sense of a classic technological fix. Rather, these technologies change humans and animals in the sense of a reciprocal constitution by significantly influencing physical abilities and thus contributing to complex human-animal-technology assemblages in the ocean. In the context of the sea, they also turn our attention towards more amphibious ways of being when certain technologies allow humans in certain ways to 'cyborg' with the sea (Haraway, 2008).

It certainly illustrates, as Gibbs has put it, 'how the making and remaking of territory beyond *terra*, and of our world, is a more-than-human project' (Gibbs, 2018: 17). And finally, the situation in La Réunion emphasises how a strong connection to the sea, as expressed by many of the surfers, bears the potential for other forms of politics, ethics and ways of knowing that increase our 'oceanic literacy' (Ingersoll, 2016: 156) and may enable us to navigate an ocean beyond borders.

References

Anderson, J. (2009). Transient convergence and relational sensibility: Beyond the modern constitution of nature. *Emotion Space and Society, 2*(2), 120–127.

Anderson, J. (2012). Relational places: The Surfed wave as assemblage and convergence. *Environment and Planning D: Society, 30*(4), 570–587.

Anderson, J. (2014). Merging with the medium? Knowing the place of the surfed wave. In J. Anderson & K. Peters (Eds.), *Water worlds: Human geographies of the ocean* (pp. 73–88). Routledge.

Barad, K. (2003). Posthumanist performativity: Toward an understanding of how matter comes to matter. *Signs: Journal of Women in Culture and Society, 28*(3), 801–831.

Barss, P. (1984). Injuries due to falling coconuts. *Journal of Trauma: Injury, Infection, and Critical Care, 24*(11), 990–991.

Barua, M. (2014). Circulating elephants: Unpacking the geographies of a cosmopolitan animal. *Transactions of the Institute of British Geographers, 39*(4), 559–573.

Benson, E. (2010). *Wired wilderness: Technologies of tracking and the making of modern wildlife*. John Hopkins University Press.

Blount, C., Pygas, D., Lincoln Smith, M. P., McPhee, D. P., & Bignell, C. (2021). Effectiveness against white sharks of the Rpela personal shark deterrent device designed for surfers. *Journal of Marine Science and Technology, 29*(4), 582–591.

Bremner, L. (2014). Fluid ontologies in the search for MH370. *Journal of the Indian Ocean Region, 11*(1), 8–29.

Brockington, D., & Igoe, J. (2006). Eviction for conservation: A global overview. *Conservation & Society, 4*(3), 424–470.

Brooks, T., Mittelmeier, R. A., Da Fonseca, G., & Gerlach, J. (2006). Global biodiversity conservation priorities. *Science, 313*(5783), 58–61.

Carter, J., & Palmer, J. (2017). Dilemmas of transgression: Ethical responses in a more-than-human world. *Cultural Geographies, 24*(2), 213–230.

Centre Sécurite Requin (CSR). (n.d.). Retrieved March 26, 2024, from https://securite-requin.re

Chapman, B. K., & McPhee, D. (2016). Global shark attack hotspots: Identifying underlying factors behind increased unprovoked shark bite incidence. *Ocean & Coastal Management, 133*, 72–84.

Cloke, P., & Perkins, H. C. (2005). Cetacean performance and tourism in Kaikourar, New Zealand. *Environment and Planning D: Society and Space, 23*(6), 903–924.

Fédération Française Surf. (2015). *French Surfing Federation on Reunion Island' shark crisis.* Retrieved March 26, 2024, from https://www.surfingfrance.com/disciplines/surf/french-surfing-federation-on-reunion-island-shark-crisis.html

Gardner, B. (2016). *Selling the Serengeti: The cultural politics of Safari Tourism.* The University of Georgia Press.

Gegan, C. (2023, June 5). Surfing next left Réunion island, even at the height of the 'shark crisis'. *The Inertia.* Retrieved March 25, 2024, from https://www.theinertia.com/surf/surfing-never-left-reunion-island-even-at-the-height-of-the-shark-crisis/

Gibbs, L. (2018). Shores: Sharks, nets and more-than-human territory in Eastern Australia. In K. Peters, P. E. Steinberg, & E. Stratford (Eds.), *Territory beyond terra* (pp. 203–219). Rowman & Littlefield.

Gibbs, L., & Warren, A. (2012, August 20). Who's hunting who? Misguided responses to shark attacks. *The Guardian.* Retrieved March 26, 2024, from https://theconversation.com/whos-hunting-who-misguided-responses-to-shark-attacks-8867

Gibbs, L., & Warren, A. (2015). Transforming shark hazard policy: Learning from ocean-users and shark encounter in Western Australia. *Marine Policy, 58*, 116–124.

Green, M., Canassin, C., & Reid, D. D. (2009). *Report into the NSW Shark Meshing (Bather Protection) Program.* NSW Department of Primary Industries, Fisheries Conservation and Aquaculture Branch.

Guinand, A., & Henchoz, G. (2016). Forbidden ocean in Réunion Island. *Ocean71 Magazine.* Retrieved June 4, 2024, from http://ocean71.com/magazine/forbidden-ocean-reunion-island-shark-attack-death-solutions/

Habiter La Réunion. (2020). *Le risque requins à la Réunion.* Retrieved March 26, 2024, from https://habiter-la-reunion.re/les-attaques-de-requins-a-la-reunion/

Haraway, D. (2008). When species meet. *Journal of Agricultural and Environmental Ethics, 21*(6), 609–611.

Helmreich, S. (2007). An anthropologist underwater: Immersive soundscapes, submarine cyborgs, and transductive ethnography. *American Ethnologist, 34*(4), 621–641.

Hinchliffe, S., & Whatmore, S. (2009). Living cities: Towards a politics of conviviality. In D. White & C. Wilbert (Eds.), *Technonatures, environments, technologies, spaces and places in the twenty-first century* (pp. 105–124). Wilfrid Laurier University Press.

Holland, K., Wetherbee, B., Lowe, C., & Meyer, C. (1999). Movements of tiger sharks (*Galeocerdo Cuvier*) in coastal Hawaiian waters. *Marine Biologe, 134*, 665–673.

Ingersoll, K. A. (2016). *Waves of knowing: A seascape epistemology*. Duke University Press.

Isaacs, J., & Otruba, A. (2019). Guest introduction: More-than-human contact zones. *Environment and Planning E: Nature and Space, 2*(4), 697–711.

Jachowski, D., Slowtow, R., & Millspaugh, J. (2014). Good virtual fences make good neighbors: Opportunities for conservation. *Animal Conservation, 17*(3), 187–196.

Knight, J. (2000). *Natural enemies: People-wildlife conflicts in anthropological perspective.* Routledge.

Lagabrielle, E., Allibert, A., Kiszka, J. J., Loiseau, N., Kilfoil, J. P., & Lemahieu, A. (2018). Environmental and anthropogenic factors affecting the increasing occurrence of shark-human interactions around a fast-developing Indian Ocean island. *Scientific Reports, 8*(3676), 1–13.

Laloë, A.-F. (2014). 'Plenty of weeds & penguins': Charting oceanic knowledge. In J. Anderson & K. Peters (Eds.), *Water worlds: Human geographies of the ocean* (pp. 39–49). Routledge.

Laloë, A.-F. (2016). *The geography of the ocean: Knowing the ocean as a space.* Routledge.

Latimer, J. (2013). Being alongside: Rethinking relations amongst different kinds. *Theory, Culture & Society, 30*(7/8), 77–104.

Lehman, J. (2018). From ships to robots: The social relations of sensing the world ocean. *Social Studies of Science, 48*(1), 57–79.

Lemahieu, A., Blaison, A., Corchelet, E., Bertrand, G., Pennober, G., & Soria, M. (2017). Human-shark interactions: The case study of Reunion Island in the south-west Indian Ocean. *Ocean & Coastal Management, 136*, 73–82.

Löe, J., & Röskraft, E. (2004). Large carnivores and Human safety: A review. *AMBIO. A Journal of the Human Environment, 33*(6), 283–288.

Lorimer, H. (2000). Guns, game and the grandee: The cultural politics of deerstalking in the Scottish Highlands. *Ecumene, 7*(4), 403–431.

Losen, B. (2023). Shark attack risk on Reunion Island: Emphasis on local media construction. *Marine Policy, 157*, 105851. https://doi.org/10.1016/j.marpol.2023.105851

Mack, J. (2011). *The sea: A cultural history.* Reaktion.

Maxwell, S., Hazen, E., Lewison, R., Dunn, D., Bailey, H., Bograd, S., Briscoe, D., Dossette, S., Hobday, A., Bennett, M., Benson, S., Caldwell, M., Costa, D., Dewar, H., Eguchi, T., Hazen, L., Kohin, S., Sippel, T., & Crowder, L. (2015). Dynamic ocean management: Defining and conceptualizing real-time management of the ocean. *Marine Policy, 58*, 42–50.

Neff, C. (2012). Australian beach safety and the politics of shark attacks. *Coastal Management, 40*(1), 88–106.

Neff, C. (2015). The Jaws effect: How movie narratives are used to influence policy responses to shark bites in Western Australia. *Australian Journal of Political Science, 50*(1), 114–127.

Nyman, E. (2019). Techno-optimism and ocean governance: New trends in maritime monitoring. *Marine Policy, 99*, 30–33.

Peters, K. (2014). Taking more-than-human geographies to sea: Ocean natures and offshore radio piracy. In J. Anderson & K. Peters (Eds.), *Water worlds: Human geographies of the ocean* (pp. 177–191). Routledge.

Picken, F., & Ferguson, T. (2014). Diving with Donna Haraway and the promise of a Blue Planet. *Environment and Planning D: Society and Space, 32*(2), 329–341.

Pratt, M. L. (1991). Arts of the contact zone. *Profession, 1991*, 33–40.

Pratt, M. L. (1992). *Imperial eyes: Travel writing and transculturation.* Routledge.

Radiator Waterwear. (2019). *SAMS shark safe visual technology.* Retrieved May 30, 2024, from https://radiator.net/pages/shark-tech

Rozwadowski, H. M. (2018). *Vast expanses: A history of the oceans.* Reaktion.

Selingo, J. (2002, June 13). How it works; When the shark doesn't bite. *The New York Times.* Retrieved March 26, 2024, from https://www.nytimes.com/2002/06/13/technology/how-it-works-when-the-shark-doesn-t-bite.html

Sharksafe Barrier. (2021). *Home.* Retrieved June 6, 2024, from https://www.sharksafesolution.com/

Sheller, M., & Urry, J. (2004). *Tourism mobilities: Places to play, places in play.* Routledge.

Shofield, D. (2024). Shark diving: The 16 best dive sites in the world. *PADI.* Retrieved March 26, 2024, from https://blog.padi.com/shark-diving-the-10-best-dive-sites-in-the-world/

Sorgue, P. P. (2017, March 10). A la Réunion, les requins déchirent le vivre ensemble. *Le Monde.* Retrieved March 26, 2024, from https://www.lemonde.fr/m-actu/article/2017/03/10/les-requins-dechirent-la-reunion_5092622_4497186.html

Squire, R. (2021). *Undersea geopolitics: Sealab, science, and the Cold War.* Rowman & Littlefield.

Steinberg, P. E. (1999). The maritime mystique: Sustainable development, capital mobility, and nostalgia in the world ocean. *Environment and Planning D: Society and Space, 17*(4), 403–426.

Sundberg, J. (2006). Conversation encounters: Transculturation in the 'contact zones' of empire. *Culture Geographies, 13*(2), 239–265.

Surmont, E. (2016). Peur sur les plages. Du "risque requin" à la "crise requin" à la Réunion. *Géoconfluences.* Retrieved March 25, 2024, from http://geoconfluences.ens-lyon.fr/informations-scientifiques/dossiers-thematiques/risques-et-societes/corpus-documentaire/peur-sur-les-plages-requins-reunion

Taglioni, F., & Guiltat, S. (2015). Le risque d'attaques de requins à La Réunion. *EchoGéo, 23*. Retrieved March 25, 2024, from https://journals.openedition.org/echogeo/14205

Taylor, C. S., & Carter, J. (2013). The agency of dolphins: Towards inter-species embassies as sites of engagement with 'significant otherness.' *Geographical Research, 51*(1), 1–10.

Thiann-Bo Morel, M. (2019). Tensions entre justice environnementale et justice social en société postcoloniale: Le cas du risque requin. *VertigO, 19*(1). Retrieved June 4, 2024, from https://id.erudit.org/iderudit/1065416ar

Todd, R., & Hynes, M. (2017). Encountering the animal: Temple Grandin, slaughterhouses and the possibilities of a different ontology. *The Sociological Review, 65*(4), 729–744.

Waitt, G., & Warren, A. (2008). Talking shit over a brew after a good session with your mates: Surfing, space and masculinity. *Australian Geographer, 39*(3), 353–365.

Wetherbee, B., Lowe, C., & Crow, G. (1994). A review of shark control in Hawaii with recommendations for future research. *Pacific Science, 48*(2), 95–115.

Weinberg, A. (1993). Chapters from the life of a technological fixer. *Minerva, 31*, 379–454.

Whatmore, S. (2014). Nature and human geography. In P. Cloke, P. Crang, & M. Goodwin (Eds.), *Introducing human geographies* (pp. 152–162). Routledge.

Whatmore, S., & Thorne, L. (2000). Elephants on the move: Spatial formations of wildlife exchange. *Environment and Planning D: Society and Space, 18*(2), 185–203.

Williamson, J. (2015, August 17). Mike Baird is right, culling sharks doesn't work – Here's what we can do instead. *The Conversation*. Retrieved March 25, 2024, from https://theconversation.com/mike-baird-is-right-culling-sharks-doesnt-work-heres-what-we-can-do-instead-46195

Wilson, H. (2017). On geography and Encounter: Bodies, borders, and difference. *Progress in Human Geography, 41*(4), 451–471.

Wilson, H. (2019). Contact zones: Multispecies scholarship through imperial eyes. *Environment and Planning E: Nature and Space, 2*(4), 712–731.

Woodroffe, R., Thirgood, S., & Rabinowitz, A. (2005). *People and wildfire: Conflict or coexistence?* Cambridge University Press.

WWF Mediterranean Marine Initiative. (2019). *Sharks in crisis: A call to action for the Mediterranean*. WWF.

Borders and Confinement in Seafarers' Realities

Maria Borovnik

Introduction

Seafarers' everyday lives are practised within structures of enclosure. These comprise the ship, and a vast array of neoliberal, regulatory and legal frameworks. Yet, the romanticised idea of a 'sailor's life' (in many films, novels and so on) often depicts seafaring mobility as a free ride across the globe, where rugged men work hard, while their ships are passing by an ever-changing exotic oceanic scenery, and where they disembark in ports that offer a good time ashore. This stereotype imagines seafarers as masculine, adventurous, rough and hardy, mobile subjects. It also assumes that there are enjoyable hours of time-off periods at shore. Today, ship crews are indeed still dominated by a mobile masculine work culture (Sampson, 2021, 2024), where women are in a strong minority (Kitada, 2021). Nonetheless, the notion of enclosure persists. Historically, sailors could be trapped in seafaring work,

M. Borovnik (✉)
Massey University, Palmerston North, New Zealand
e-mail: M.Borovnik@massey.ac.nz

K. Peters and J. Turner (eds.), *Ocean Governance (Beyond) Borders*,
https://doi.org/10.1007/978-3-031-71322-4_10

press-ganged into time at sea (Ogborn, 2008). Today, forced or enslaved labour is known to continue to exist off-shore (Urbina, 2019). But even 'regular' working conditions are enclosing seafarers. Indeed, in contemporary times of capitalist-driven globalisation, free periods on ships and options to spend time ashore are severely restricted. Embedded in the ever-competitive maritime trading system, seafarers are conditioned by both the apparent freedoms of crossing a changing ocean-scape and the un-freedoms of the restraints of work-life worlds on ships, which are determined by deep-seated postcolonial structures that continue under a neoliberal market system. For the shipping industry to operate well its essential needs are of uninterrupted trade flows that promptly fulfil contract arrangements. Seafarers are expected to perform towards the end goal of securing global trade needs. Within these expectations, the hierarchical ship-setting controls all work. Seafarers must be readily available for 24/7 activities that subordinate the ship-based workforce under the demands of global trade.

The challenges this uneven system continues to generate reached a peak during 2020 after the World Health Organisation declared the spread of COVID-19 as a global pandemic (WHO, 2023). Starting in March 2020, in response to the pandemic, most countries across the globe closed their borders. These global lockdowns forced ships to a temporary standstill. Unable to access shore, more than 400,000 seafarers were left stranded (IMO, 2022). This crisis was recognised as a humanitarian issue. It had 'turned ships into "floating prisons"' (Luchenko & Georgiievskyi, 2021: 8), where seafarers were unable to meet their basic needs or have access to health care. At the same time, it was not possible for companies to replace those aboard that had reached the end of their contracts with others, affecting seafarers who were waiting ashore (Slišković, 2020; Banta & Pratt, 2023). Astonishingly, it was this crew change crisis that highlighted how seafarers had never been recognised as essential workers, although they were offering essential services by transporting products from places of production to sites of consumption for centuries.

This chapter illustrates how seafarers' realities are influenced by factors and decisions outside their power, predisposed by uneven global structures (Harvey, 2015), and how these factors and decisions had immobilising effects for seafarers servicing the global economy. These unequal global transport conditions (Sheller, 2018a), and the marginalised placing of seafarers within uneven politics of movement (Kotef, 2015), are a core recognition in this chapter. The main argument throughout is based on Sheller's concept of mobility (in)justice and explores the oxymoron of seafarers as both drivers (agents) and servants (subordinates) of global trade simultaneously. Seafarers must continuingly negotiate their agency within these systems. The chapter expands Sheller's concept in the next section and considers mobility (in)justice in relation to oceanic bordering. It then turns to the border closures during the COVID-19 pandemic to reflect on how pandemic governance enacted new forms of ocean governance on seafaring mobilities.

A Ship Culture of (Im)Mobilities Through the Lens of Mobility Justice

Almost 90% of global trade is transported by ship, involving a complex network along ship and shore (ICS, 2021). Operating within the 'forgotten space' of a capitalist global system, which functions by 'idealising the erasure of distance', as Steinberg (2015: 40) phrased it, the demands of the shipping industry normalise seafarers as mobile beings. Without seafarers, globalisation would not be possible (Borovnik, 2011b, 2022; Markkula, 2021). And yet, despite their vital role in the transport sector, seafarers are dealing with precarious issues that are (re)produced by 'dominant mobile regimes' (Sheller, 2023: 435). Sheller (2018a, 2018b, 2023) explains that (im)mobilities are intrinsically entwined with unequal power relations, which have their roots in different forms of racial and imperial-colonial informed capitalist structures. These mobility regimes continue (re)producing unequal networks, relations and 'regimes' on different scales.

Recognising the uneven power relations implicated in all movements, Sheller (2018a, 2018b, 2023) identifies mobility and immobility as intrinsically political. The politics of mobility are connected with the complexities of social relations (Cresswell, 2010), which are fluid (Urry, 2008: 14), (re)produce social inequalities and 'reinforce social difference' (Cook & Butz, 2018: 5). Sheller (2018b: 32) explains that a mobility justice approach focuses on the multiple intersections, refractions and intensifications of 'cross-cutting problems of the politics of uneven (im)mobilities' and resulting injustices. A mobility justice approach addresses these issues on different scales by using a mobile ontology—which expands our thoughts beyond a containerised notion of mobilities that is determined by the space within which they operate (Sheller, 2018b; Verlinghieri & Schwanen, 2020). By centring on mobilities, and all the complexities involved, and paying attention to the intersection of postcolonial insights to global (im)mobilities, a mobility justice approach offers one common framework that connects with other justice approaches (Sheller, 2018a, 2020) and intersects these mobility injustices with embodied racial and gendered experiences (Sheller, 2020: 33). In this sense, mobility justice serves as a multifaceted, intersectional framework around the embodied and entangled power relations that play out on both micro- and macro-scales (Harada, 2023). These include (or exclude) humans, resources, things, information and ideas, that enable or disable access, or the ability to engage in movement (Cook & Butz, 2020: xx; Sheller, 2023: 436). Shipping as a mobile assemblage of networks can serve as a vivid example that shows how these entangled relations and social processes play out. It is also useful to consider Soja's argument that 'no social process takes place uniformly over space; there will always be some unevenness in the geographies we produce' and therefore, inequalities arise (2010: 71). A mobility justice framework seeks to address these inequalities, which include the continuing 'practices of borders and borderings' that 'have contributed to this twenty-first century silencing discourse of global coloniality' (Wemyss, 2023: 1).

Such bordering issues are a continuation of colonial structures, fostered by enclosures that result in 'uneven access to movement and mobility space' (Sheller, 2023: 435), including control mechanisms and devices that constrain or restrict the movement of people and goods (Gallez, 2023: 38). There are perhaps two angles from which we can view restrictive bordering practices in a seafaring context: first, the limited ability to access ships and seafaring work; and second, the regulated (in)ability to access shore during contracts on ships. Both draw on immigration and visa status and both use border control mechanisms. Both are influenced by 'top-down', 'power-over' relationships, where some countries produce visas that allow more access across borders than others. Both also rely on nation states' compliances with international agreements, such as human rights declarations and the Maritime Labour Convention (MLC). These underlying structures are uneven and determine 'who can travel, when, where, and how', as is acknowledged in mobilities research (Adey et al., 2021: 2).

As mentioned earlier, this unequal access to movement is influenced by racial, gendered and postcolonial systems. To address the underlying power structures, Cresswell stresses the entanglements that 'constellate' mobilities within their social relations and histories as they determine how movement is represented and practised (2010: 19). Indeed, elaborating on complex mobilities power systems or 'constellations', Cresswell (2010: 27, drawing on Torpey, 2000) shows that disparities that regulate, define and control 'legitimate' movement also began well before the establishment of nation states borders and the use of passports. Mobility norms often rest on narratives of liberty, with mobility associated with being progressive, and with transgressive mobilities linked to those moving in ways other than in what would be socially deemed 'acceptable' (i.e. the movement of vagrants, see Cresswell, 2001, 2016). The equation of mobility as a resource of liberty for those moving within accepted parameters (i.e. for tourism) has fitted well with those who are financially affluent and able to move at their leisure, but not so much with those who are struggling economically, and who are either restricted or forced to move. As Gallez (2023: 40) notes mobility can be either 'claimed as a right or denounced as oppression'. These discriminations have continued with the demands of labour mobilities that allow loose contract commitments, which are evident in freight and shipping companies, and demand extreme availabilities and flexibilities by seafarers. Ironically, within these requirements, seafarers' freedom of movement has been increasingly restricted.

Limiting seafarers' mobilities is linked with the necessity for efficiency in the shipping sector. Cost-saving practices by shipping companies use 'flagging out' procedures, which by the 1980s have become mainstream (Alderton et al., 2004).[1] Shipping companies may use a flag in agreement with foreign nations that allow relatively lenient employment standards. Using a foreign flag, what is also labelled as a 'flag of convenience', enables companies to employ crews outside their own nation state. This practice has resulted in recruiting multinational ship crews, of which a majority are from the Global South (Baum-Talmor, 2021). Seafarers from these areas appreciate the employment opportunities that the global shipping industry offers. Chances presented by the international maritime sector allow much needed financial support for seafarers' families, as jobs in home countries can be often difficult to obtain and wages in international seafaring employment are usually higher than local earnings (Baum-Talmor, 2021; Markkula, 2021). Employment under flags of convenience, however, comes with a cost and seafarers must endure contracts that offer less security and longer time-periods away from home. Indeed, the freedom of movement that is linked with access to employment for many is also restricted by regulations and recruitment requirements (Markkula, 2021). Sheller (2018b: 27), referring to both Amartya Sen and Martha Nussbaum, contemplates the notion of 'freedom of movement as a capability'. The capability to move makes it possible for people to access goods and services (and in this case, work); but then again, and depending on the external or social structures, a person's functioning may be constrained. In a seafarer context, there are obvious ambiguities. Seafarers from the Global South can access work in the global shipping sector. Yet access within the ship space itself is hierarchically ordered. Access to shore depends on status and visa permits. And even though basic needs are covered during employment on ships, it is often difficult to attain goods beyond basic needs or even to access health services. Therefore, it is not only important to emphasise freedom of access to mobility, but also to look at equality of movement (Harada, 2023: 425), and the quality of a seafarers' functioning. Foreign flags allow the shipping sector to work at lower costs and with higher

[1] Using foreign flags had also been practiced during war times early on in the nineteenth century in order to avoid being entangled in conflict, but from the 1980s onwards this practice started to gain strong momentum.

profitability.[2] To stay afloat in an extremely competitive world market such efficiency measures are necessary for companies (Borovnik, 2011b). And yet, these longer contracts can place seafarers within tiresome and monotonous situations (Borovnik, 2011a), where there is little escape from the confinements of the relatively narrow ship space.

During economic recession (for example during 2008 and 2009),[3] efficiency measures also witnessed the shrinkage of crews, keeping crew numbers at a minimum. This has further restricted options for time off and relaxation. While ships are unloading, loading or at bunker, the duration and frequency of watch-keeping has increased. The extraordinary stressful and physically demanding work that seafarers conduct daily would normally require rest-time—except that relaxation is now a rare reality for seafarers. When anchored at port, duties today take up so much time that shore leave is limited or impossible. With the introduction of containerships, ports have become specialised with high-capacity technology to offer efficient and fast turnaround. The combined effects of remoteness of high-functioning containership ports, the abbreviated opportunities for shore leave and increased security and port control measures have jointly enacted new temporal and regulatory borders causing seafarers to stay aboard their vessels. In these current circumstances, some might never go ashore during their entire time of contract, as their duties keep them so busy that they need any free minute for rest. And so, for seafarers today, freedom of movement is severely limited. It is limited by a multiplicity of mobility bordering strategies, both tangible and more abstract, from the specific regulatory landscape of ship flagging, ports and border controls, to the hazy, yet very real impacts of the just-in-time economy, and market dynamics.

Yet freedom of movement has been recognised as a human right in the Universal Declaration of Human Rights (UDHR), specifically the International Covenant on Civil and Political Rights (ICCPR) and

[2] One example provided by De Beukelaer (2021) during the COVID-19 pandemic was Panama's flag allowing first up to 14 months, then even to 17 months extension of on-board time for seafarers.

[3] Recession was ongoing in 2009, when I spent time on a containership. It was very kind of the shipping company to allow me to spend a month with minimum crew towards the end of 2009 and January 2010.

Covenant on Economic, Social and Cultural Rights (ICESCR) (UN, n.d.). However, the actual liberalist idea of freedom must be approached with some caution. Schubert (2021) in his analysis on Foucault's interpretation of freedom as an inadvertent component of power explains neoliberal power as repressive. Freedom can subtly subject individuals to 'power-over' predicaments. By 'creating incentives to become free and even resistant' against 'inefficient structures', individuals are steered towards progress and innovation, useful for neoliberal, market-driven agendas, simultaneously keeping them trapped within systems of control (2021: 643). As such, the free market continues to shape uneven spatial, racial and gender/sexual relations (Sheller, 2018b: 28). Within neoliberal subjectification freedom appears murky and resistance against neoliberal government is not actually possible, which makes neoliberal-driven freedom questionable, unjust and undesirable (Schubert, 2021: 643).

The freedom promoted by neoliberal global free market systems actually immobilises seafarers, who, as the essential drivers of these systems, are simultaneously subjected to and enclosed within regulating structures that are determined by external decision-makers (Borovnik, 2022; Sampson, 2024). Neoliberally driven globalisation, therefore, is not only facilitated by seafarers—it drives their subjectification and creates mobility borders for them. Seafarers, while operating the vast majority of global trade, are instantaneously enclosed in repetitive loops of (im)mobilities. Sheller (2020) has suggested focusing on the different scales of mobility regimes, which is valuable in the seafarers' context, where the global neoliberal mobilities regime has direct effects on the embodied scale. In this sense the romanticised idea of seafarers as extraordinarily tough and hardy is taken advantage of, their resilience is drawn on to continue the unequal economic regimes and social relations in which they operate. As seafarers enter the global job market to provide for their families at home, they will simultaneously take on the disadvantage of arduous hard work on ever-moving ships, in which they share a comprised living space over prolonged time with others. These tough circumstances become even more gruelling under extenuating conditions, such as when merchant ships have to cross piracy or war conflicted zones or as was recently the case with the prolonged crew change crisis that started at the peak of the COVID-19 pandemic, as the next section will draw attention to.

Border Closures and (Un)Freedoms at Shore and Sea

A burst of global acknowledgement dealt with the immobilising effects for seafarers soon after the COVID-19 pandemic peaked in 2020, which resulted in a crew change crisis that affected 400,000 seafarers who were stuck on ships (Bailey et al., 2021; ICS, 2021; IMO, 2021a–d; UNCTAD, 2020), and a similarly high number of seafarers stuck at home (De Beukelaer, 2021; IMO, 2021d). In October 2020, at the peak of the crisis, 43 states and territories closed their borders for crew change and, during 2021, 24 states still had their borders closed (De Beukelaer, 2021: 2). This humanitarian crisis distinctively highlighted the ongoing—and overlooked—issues of uneven global transport conditions seafarers were facing while steadily and 'tirelessly' continuing to support 'the often invisible global logistics chain' (UN Secretary-General António Guterres in UNRCCA, 2020), well beyond their normal contracts and with even more prohibitive restrictions to shore access. For many seafarers an end to their trapped situation on vessels during the pandemic felt dreadfully uncertain. They felt they were serving 'an unwanted prison sentence' (SeafarerHelp, 2020: n.p.), unable to leave their ships.

Identifying the toll the crew change crisis had on seafarers, in September 2020, the United Nations (UN) Secretary-General António Guterres, stressed that seafarers needed recognition as key workers to 'ensure safe crew changes' and 'allowing stranded seafarers to be repatriated and others to join ships' (UNRCCA, 2020). Led by a Seafarers Crisis Action Team (SCAT), the international community launched several initiatives towards negotiating seafarers' rights and reminding countries of their obligations to provide basic services to crews, including port access, medical treatment and the ability to leave the constraints of ship space. Despite these efforts, the crisis continued to worsen throughout 2021, when globally, 'seafarers stuck aboard ships beyond the expiry of their contracts rose from 5.8% in May 2021 to 8.8% in July

2021' (Shipping Australia, 2021: n.p.). In 2022, there were still 200,000 (or approximately 2.7% of the global seafarer workforce) stranded (ILO, IMO, UNCTAD and WHO, 2022; Schuler, 2023). Intensifying this situation, the war in Ukraine had added to this crisis, necessitating the replacement of about 2000 officers from Russia and Ukraine during the same year (Foley, 2022; Human Rights at Sea, 2022; Safety4sea, 2022). Considering these complexities, the IMO implemented a resolution to evacuate seafarers from war zones bordering the Black Sea and the Sea of Azof (UNCTAD, 2022).

On an international scale, dominated by these COVID-19 preventative border closures, the response and call for safe crew changes was cooperative, mostly driven by the urgency to continue the flow of global supply chains and smooth operations for shipping companies (De Beukelaer, 2021). Yet, the crisis revealed 'governance mechanisms at different scales ... demonstrating problems of coordination, uneven capacities and claims of authority', which affected 'countries and communities differently, due to different levels of preparedness of their health, social security systems and infrastructure capacity' (Blanco & Rosales, 2020, in Doumbia-Henry, 2020: 5). De Beukelaer (2021: 2) observed that even when borders had begun to open there were numerous obstacles that kept seafarers immobilised, where 'shipping companies were faced with costly crew changes, due to the limited availability of flights through much of 2020', and where some chartered vessels began to be 'using 'no crew change' clauses, particularly with bulk carriers, for fear that crew change may lead to bringing COVID-19 aboard'. In De Beukelaer's view, those practices served as an 'excuse to violate MLC, 2006 regulations' (2021: 2). During 2021, crew changes remained difficult but were possible by using alternative routes, and obliging with long quarantine arrangements for crews, and high flight costs because of irregular flight availabilities (De Beukelaer, 2021). Moreover, Devereux and Wadsworth (2022) have highlighted how seafarers, depending on their employment situation and length of contract time, may not necessarily have had access to social protection and those in precarious situations were affected by poorer health and financial adversity. In other words, the COVID-19 crew change crisis revealed challenges on the global, national and embodied scale of Sheller's mobility justice regime, where states operated from different vantage points while dealing with the social and economic consequences of this crisis, where the shipping industry continued finding strategies to balancing their international crewing agreements with economic costs, and where seafarers were on the bottom of everybody's priority lists, despite their important role as drivers of the global market.

On these different scales, the consequences of continuing interruptions were felt by suppliers and consumers who were facing disruptions and delays. The world started realising the crucial role of seafarers as key workers—transporting consumer goods across the globe—when these commodities did not arrive (Markkula, 2021; Munoz, 2022). In Graham's (2022: n.p.) opinion, seafarers were so obviously essential to global trade that it seemed ironic how consumers ashore only began noticing them when a shortage of goods was pending, and that seafarers 'are expected to be safety, security and humanitarian operatives while at sea, and trusted to bring cargo to ports; yet, ironically, in many ways they are thought of and treated as threats at borders'. With the advocacy by the UN Secretary-General, the analysis by SCAT, and the media highlighting their situation, seafarers were, at last, officially supported by the Neptun Declaration, which was signed by 850 organisations towards facilitating crew changes (Global Maritime Forum, 2023). Despite these efforts, authors, such as Banta and Pratt (2023), highlight the extended, ongoing immobilities and precarities of seafarers, who remain(ed) undervalued (ILO, 2023; Munoz, 2022), and whose voices continue(d) to be silenced, during and beyond the pandemic (Wemyss, 2023). Despite everybody at least temporarily acknowledging the key role of seafarers, Graham (2022) noted not only a hesitancy of states in signing the essential workers designation; there was also reluctance with providing vaccinations to seafarers in ports. When states started opening borders in 2022, and began 'to return to normal', Graham found that 'the special treatment that some countries afforded seafarers [to be] waning' (2022: n.p.), which was, in her observation, quite in contrast to the aviation industry, where workers were treated with higher priority.

The COVID-19 pandemic has strongly underlined the importance of looking more deeply into the occupational risks that being trapped on ships implicate. The International Chamber of Shipping (ICS) found in their Seafarer Shore Leave Principles report published in April 2022 that 'governmental restrictions imposed to reduce risk of potential infection has had severe consequences on the ability of seafarers to take shore leave when in port' (ICS, 2022: 2). A number of publications have underlined the impacts on mental, physical and social well-being for those who were stuck on ships (Brooks & Greenberg, 2022; Jonglertmontree et al., 2023; Lucas et al., 2021; Slišković, 2020). In her profound study, Ana Slišković (2020) conducted a qualitative analysis with 750 COVID-19 affected seafarers of different nationalities and found that seafarers felt exhausted, deserted and forgotten, as these example responses show: 'We are already tired and our body and mind has reached its limits, but the shipping company doesn't understand us. We are not safe here in the vessel'. Others said: 'Feel like in jail!' or 'I feel abandoned by society; that I am good only like a slave but not as part of human community' or 'They forgot us' (Slišković, 2020: 804). Human Rights at Sea had conducted a study with I-Kiribati seafarers. One had joined his vessel in October 2020, saying that when joining the ship, he had to keep in mind that because of the pandemic:

> ...we don't have any place to go ashore anymore as seafarers, because every country has restricted seafarers from having shore leave. You're still working every day from morning till evening, and it's like you are in a prison. Getting off from the ship is always the first thing on our mind that we're always waiting for. (Williams & Nicholls, 2021: 15)

Referring to the International Transport Federation, De Beukelaer (2021) noted that seafarers fear being blacklisted (losing job opportunities) if they would speak up about ship safety being impacted because of physical or mental health issues. While many seafarers stranded on ships have felt abandoned and overlooked by society, Slišković (2020) also noted that those who were stuck at home, waiting for new job opportunities and not able to travel for their usual crew turnover were also stressed about their economic well-being and not being able to supply their families with financial means.

The following two examples will highlight some details about the effects of border closures and limited access to shore on seafarers during the COVID-19 crisis. The first example looks at New Zealand seafarers. New Zealand does not have its own merchant fleet and seafarers are instead working for international shipping companies. Some are employed in the cruise ship sector, others on floating rigs, or hired by yachting charterers. The second example considers seafarers from the Republic of Kiribati, an Island nation state in the Central Pacific that, despite this remote location, has had well-established contracts with international shipping companies specialised in merchant trade since the late 1960s, and from the mid-1980s also in the fishing industry.

New Zealand's Managed Isolation and Quarantine System Hindering Repatriation and Flexible Crew Turnaround

In April 2021, initiated by their 'seafarers are essential' campaign (Change.org, 2021), a group of New Zealand seafarers started a public Facebook group (facebook.com/NZSeafarers/ [2021]),[4] encouraging others to sign up to their campaign and to share their stories on Facebook. A media release by Change.org explained that, despite New Zealand's support of designating seafarers as key workers and their commitment to the Maritime Labour Convention (MLC) 2006, seafarers were not on the list of persons prioritised for vaccinations by New Zealand port authorities. This decision affected seafarers' options to take on jobs that required vaccines. Without these vaccines, seafarers were not able 'to transit in and out of ports', nor to return easily to their families once they signed off (NZSeafarers, 2021: n.p.). Vaccine priorities were not the only issue: the most burning problem, according to the Change.org campaign, was the difficulties of accessing Managed Isolation and Quarantine (MIQ) placing. This MIQ system was introduced on 26 March 2020, immediately after New Zealand had closed its borders. The Maritime Border Order required this MIQ time for anyone when entering the country until 2 May 2022, when the Order

[4] Just before the actual publication of this chapter, this Facebook page has been closed for public viewing.

was removed[5] (MBIE, n.d.). One of the spokespersons of NZSeafarers described on 12 April 2021 that there was an urgency for the government, as a signatory of the IMO designation, to recognise seafarers as key workers, to put their support to seafarers into practice and treat them equal to other essential workers, such as air crew members, in prioritising MIQ spaces for them. This was urgent, the spokesperson exclaimed, because many seafarers were in prolonged shipboard or precarious transitory situations (RNZ, 2021).

Yet the difficulty for seafarers to secure MIQ for themselves symbolises the abovementioned constraints of ship work. There is extremely limited internet on ships, if at all. Internet access is usually only available when ships are nearing ports. Using the internet is also normally volatile and quite expensive. These uncertainties of internet use have constricted seafarers' probabilities of accessing MIQ spaces, as they had to compete with hundreds of others queuing up for the small numbers of MIQ placings available. Seafarers were disadvantaged when internet coverage suddenly ran out and they had to start applying again from the back of the MIQ queue. As a consequence, many had to extend their time on ships or had to disembark in countries without an appropriate visa, in which case they had to wait for their return in quarantined environments, often grouped with seafarers from other nationalities in hotels and youth hostels, with very limited access to the outside. Even though many employers were understanding for a while, some decided not to renew contracts as a consequence of these uncertainties (Personal Conversations, November 2023). The following two paraphrased narratives,[6] chosen as typical examples, were uploaded by seafarers on the NZSeafarers Facebook page.

One seafarer had left New Zealand in early October 2020 expecting to be home for Christmas. By the time of posting on Facebook he had hoped to return home by early June 2021. He explained how much he missed his family and many friends who were quite supportive during the time he had been 'basically locked out' of New Zealand. He then

[5] On 1st July 2023 national quarantine and isolation requirements were transferred back to the national health system.

[6] For privacy reasons I have removed names of seafarers and have paraphrased their Facebook entries.

explains how much he regretted: not being home when his wife was in hospital after a car accident, not being able to visit his elderly mother, who was placed in dementia care, and not being available for all 'the important people in my life' (posted 11 April 2021). Another seafarer had also left home in October 2020 and was due home towards the end of July 2021. He wrote that he missed being involved in his family's day-to-day lives, such as making coffee for his wife in the morning, taking the kids to school, picking them up in the afternoon and taking them to their sports. When he left, he had expected to be away for about three months which would have included the isolation periods (posted 15 April 2021). These personal accounts show that New Zealand seafarers (as well as their families) suffered with the extended time away from their loved ones, as they were not used to prolong times off-shore.

During New Zealand's border closures, seafarers were misunderstood by the New Zealand COVID-19 team at the Ministry of Business, Innovation and Employment (MBIE, n. d.), which did not allow the flexibility that would have been required to fairly accommodating their needs. MBIE had not considered the difficulties that the constraints and mobilities of the ship space presented, of which the volatile and limited internet access was only one problem. The other occupational challenge impacting on MIQ access was that seafarers could never predict exactly at what time they would arrive at a particular port, because the nature of shipping depends on weather conditions, availability of pilots at ports and waiting times that delay cargo turnarounds or other unexpected events that may delay a ship's journey. As a consequence, seafarers might arrive in their home countries several days later than anticipated. Flight bookings are kept flexible to accommodate these frequently occurring delays. During the COVID-19 restrictions, MIQ space allocation needed to be organised using an adaptable approach and be flexible in order to assist these occupational circumstances. Yet, under great pressure to accommodate the multiple requests by travelling citizens returning home, the Ministry decided *against* the option to keep MIQ flexible for seafarers. Instead, others were prioritised. MBIE did not realise how these decisions had contradicted New Zealand government's signature in support of seafarers as essential workers designation nor how the global trade chain was impacted. On 12 August 2021—within only three months—the Change.org campaign on 'Seafarers ARE essential' had already received 3006 signatures; on this same day the NZSeafarers Facebook page of their campaign promotion had recorded 40,384 views and 719 shares. Here, governing the pandemic led to the further governing of seafarer mobilities through detrimental bordering practices, exacerbated by the operation of MIQ.

Kiribati's Zero COVID-19 Policy Affecting Repatriation with the Consequence of Loss of Employers

Kiribati is a Pacific Island nation of 32 atolls and one high-lying island, which is dispersed over 3.5 million square kilometres across the central Pacific ocean, with a land area of only 810.5 square kilometres and a population of approximately 215,000 (CIA, 2023). Until the South Pacific Marine Services (SPMS)—the employer of Kiribati's merchant seafarers—closed their offices at the beginning of 2022, merchant seafarers were the main contributors of remittances received by Kiribati for more than 50 years. Remittances in 2019 contributed to 11.3% of GDP, but this number had fallen to 6.7% in 2022 (The World Bank, 2023). This financial drop reflects the impacts of the crew change crisis on I-Kiribati seafarers between March 2020 and December 2021. Kiribati closed their border in late-March 2020 and remained free of COVID-19 until August 2021. With consideration of the vulnerability of the island population facing insufficient capacity to deal with an outbreak of the pandemic, the country had decided to keep their borders closed during 2020 and 2021, facilitating repatriation selectively (Borovnik, 2024; Borovnik et al., 2021; Williams & Nicholls, 2021). As a consequence, many seafarers remained stranded overseas, some for up to two years, and without appropriate visa they were bound to their hotel or youth hostel accommodation with limited mobility. Shipping companies of the SPMS conglomerate took care of their basic needs and supplied vaccinations (Borovnik et al., 2021), urging the Kiribati government to facilitate transport for seafarers to return home.

In September 2021, Human Rights at Sea, published an independent case review, noting that

> over 250 I-Kiribati seafarers have been discharged from their ships, isolated, and then left waiting across several countries, including Australia, pending repatriation. Most have completed their legal contracts with many having been away from their homes for over two years with periods up to eight months living in a local hotel somewhere around the globe, far from their homes and families. Given the average contract under the Maritime Labour Convention (MLC) 2006 of a Kiribati seafarer is 11 months, this additional delay in their returning home is having a significant physical and emotional impact on the seafarer, their families and their communities, with no end in sight. (Williams & Nicholls, 2021: 4)

Kiribati had returned in total 210 seafarers since the outbreak of the COVID-19 pandemic (Williams & Nicholls, 2021), but in April 2021 a COVID-19 outbreak in Fiji caused a lockdown and standstill of air travel out of Fiji. Consequently, 159 I-Kiribati were stranded in a small hotel in Nadi, while another 105 were waiting for repatriation in Australia, South Korea, Indonesia and Germany (Borovnik et al., 2021). Many of these men had already been spending months in lockdown waiting in diverse locations before reaching these last destinations. For example, one person told me that he was stranded in a hotel in Spain in 2020 for about a month, where he spent his time with a group of seafarers of all sorts of different backgrounds. They were only allowed to go for short walks outside the hotel. After this time, he joined a ship to Australia and spent approximately three months in Brisbane before he joined the group who had to endure the lockdown in Fiji. Other seafarers told me that they had been either in Denmark or Germany for several months before they landed in Fiji.[7] Seafarers were feeling low and concerned about their wives and family:

[7] During two visits to Kiribati in November 2022 and June 2023, I had the opportunity to talk to several seafarers. Official research to understand the different perspectives involved during the COVID-19 border closure, however, will yet to be undertaken and my research application is in progress.

> My son's birthday is coming in this month. I hope he will get maybe only one cake, but that's all. No presents or anything like this. If I was there already, I would work a little bit more, like fishing and getting some other kind of food that we can get, and then we could use some money to buy the kids presents for their birthday. But now, those things will slip away a little bit. (Williams & Nicholls, 2021: 16)
> We have kids who go to school, some of our seamen lost their wives, they went for another husband... it's a very sad thing, sitting down here doing nothing, earning nothing, and all the time you've occupied with problems. (Wasuka, 2021: n.p.)

In August 2021, the two largest shipping companies of the SPMS conglomerate,[8] Hamburg Süd and Maersk, notified their I-Kiribati employees that they had left the SPMS because of the prolonged difficulties with repatriation. As a consequence, the SPMS 'terminated their contract with the Kiribati government' (Williams & Nicholls, 2021: 11). Kiribati had extended their border closure until 31 December 2021 and seafarers waiting to be repatriated were now not only affected by their prolonged separation from families, but they also had to deal with the ongoing economic uncertainty of losing future employment:

> But [I also worry about] when the companies change their minds to stop our salaries, that I will have to stay in Brisbane or in Fiji until further notice. Our country is still in lockdown, and we still don't know when they will open the border for us. The biggest thing I worry about is my family, and if someone will feed them when my salary stops. I am worried about my kids' education, and if someone will support them because their father is far away from them. (Williams & Nicholls, 2021: 17)

The Human Rights at Sea study has contained many more statements of the difficult thoughts that have been occupying I-Kiribati seafarers' minds. Unfortunately, reality struck when the SPMS offices closed, and

[8] The South Pacific Marine Services was established in 1969, after Hamburg Süd had seen opportunities in Kiribati, and was first a conglomerate of British and German companies; before the disestablishment at the end of 2021 (after 52 years), the Danish Maersk, which at this time had also owned Hamburg Süd, and five German companies were part of the SPMS.

approximately between 600 and 800 seafarers have had to look for alternative employment (Personal communication in Kiribati, June 2023). It is striking to compare these two examples, the situations of I-Kiribati with those from New Zealand. Family- and financial-related concerns were at the forefront for all seafarers who unexpectedly had to be separated from their homes over longer times. Yet, the effects of financial withdrawal for the I-Kiribati seafarers created further border complexities and resulted in much more prolonged and severe times of 'unwanted imprisonment' during lockdown, than for most other seafarers.

Conclusion

A sailor went to sea, sea, sea,
To see what he could see, see, see
But all that he could see, see, see
Was the bottom of the deep blue, sea, sea, sea.

This playful nursery rhyme horrifically became a reality for more than 400,000 seafarers who were deeply impacted by the crew change crisis. This humanitarian crisis was triggered by world-wide border closures as a consequence of the COVID-19 pandemic. Seafarers had to endure what seemed an unending captivity on ships and extremely constraint living conditions. Ships are vulnerable spaces exposed to weather and sea currents and under normal procedures, seafaring is regarded as one of the most dangerous occupations (Sampson, 2024). It is also an occupation that borders mobile workforces considerably through a variety of neoliberal, regulatory and legal apparatus. Yet, the pandemic exacerbated these already constraining and enclosing conditions (Borovnik, 2024). Where shore leave was difficult to attain under usual operations, it became impossible with locked down border regulations. During the pandemic, while stuck aboard ships, not only were crews unable to access shore they could also not transit or be repatriated. Consequently, many seafarers felt deserted by wider society, including some shipping companies, and even in some cases their own nation states. This chapter has unpacked the mobility injustices that were exacerbated by the pandemic. The already existing uneven power structures that had placed seafarers on the bottom of hierarchies, and at the periphery of global attention, became even more obvious with this global humanitarian crisis.

The COVID-19-related crew change crisis worsened symptoms of what has been the norm for seafarers who are operating in the 'forgotten spaces' of global capitalism (Steinberg, 2015: 40), where they are entangled in systems of social and economic inequalities, which are intensified by uneven global mobilities injustices (Sheller, 2020: 32). Bordering practices—including decisions that were made on global and national scale during the pandemic (from the MIQ in New Zealand to pandemic border closures in Kiribati)—continue to selectively prioritise who will be allowed to enter shores, and whose repatriation will be prioritised. In this sense, the pandemic has resulted in a continuation of objectifying and silencing processes for seafarers, maintaining their placing on the bottom of hierarchies with enormous impacts on their physical, mental, social and economic well-being (De Beukelaer, 2021; Slišković, 2020; Wemyss, 2023; Williams & Nicholls, 2021). This conclusion is surprising when there has been much talk about seafarers' role as essential workers at least during the pandemic. Yet, only 63 nations have signed the designation that acknowledged seafarers as 'key workers' (Safety4Sea, 2022). Graham (2022) had compared the treatment of seafarers with those of the higher prioritised aviation crews, which was also pointed out by NZSeafarers on Facebook. It must be asked why seafarers feel as though they are threats, *why they are constructed as threats* or, what one person told me in November 2022, that they are discriminated as 'very dangerous people'—because, their highly mobile status, in spite of their ship-based enclosure and numerous exposures to control and testing, had them regarded as people who carry infectious diseases.

This chapter started with questioning the idea of a 'sailor's life', with its reputation as a life of freedom and high mobility. This chapter has shown how the realities of seafarers are determined by the demands of a free global market, which is repressive and immobilising and subjectifies seafarers under the demands of trade—and consumers. During the COVID-19 pandemic where hundreds of thousands of seafarers were stuck, both on ships and at home, the resilience of seafarers was applauded. As Secretary-General António Guterres outlined, they had 'continued to tirelessly support the often invisible logistics chain', even though 'exhausted [and] away from their families and loved ones' (UNCTAD, 2020: n.p.). Despite this appreciation of seafarers as continuing global market workforce, they had to face even more restrictive bordering practices, and efforts to aid seafarers during the pandemic were falling painfully short. Choosing to continue their lives at sea is usually the only option for a majority of global seafarers, and especially those who come from backgrounds where remittances contribute significantly to a country's national income and the well-being of families left behind. In contrast to the subjectification (or objectification) of seafarers as mere trade facilitators, the narratives in this chapter have brought a human face to maritime worlds. While shipping companies and nation states were wrangling with economic, political and civil responsibilities—closing borders and enacting a governance of the pandemic that was also a governance of the oceans—these human beings were left worried and despaired, felt encaged and without hope.

References

Adey, P., Hannam, K., Sheller, M., & Tyfield, D. (2021). Special Issue Introduction: Pandemic (im)mobilities. *Mobilities, 16*(1), 1–19.

Alderton, T., Bloor, M., & Kahvesi E. (2004). *The global seafarer. Living and working conditions in a globalised industry.* International Labour Office.

Bailey R., Borovnik M., & Bedford C. (2021, June 3). Stranded seafarers: An unfolding humanitarian crisis. *DevPolicyBlog*. Retrieved May 23, 2024, from https://devpolicy.org/seafarers-in-a-covid-world-20210603/?utm_source=rss&utm_medium=rss&utm_campaign=seafarers-in-a-covid-world-20210603

Banta, V., & Pratt, G. (2023). Immobilised by the pandemic: Filipino domestic workers and seafarers in the time of COVID-19. *Transactions of the Institute of British Geographers, 48*(3), 556–570.

Baum-Talmor, P. (2021). Careers at sea: Exploring seafarer motivations and aspirations. In V. O. Gekara & H. Sampson (Eds.), *The world of the seafarer:*

Qualitative accounts of working in the global shipping industry (pp. 51–62). Springer.

Blanco, M. L., & Rosales A. (2020). Global governance and COVID-19: The implications of fragmentation and inequality, *E-International Relations*. Retrieved May 23, 2024, from https://www.e-ir.info/2020/05/06/global-governance-and-covid-19-the-implications-of-fragmentation-and-inequality/

Borovnik, M. (2011a). Occupational health and safety of merchant seafarers from Kiribati and Tuvalu. *Asia Pacific Viewpoint, 52*(3), 333–346.

Borovnik, M. (2011b). The mobilities, immobilities and moorings of work-life on cargo ships. *SITES, 9*(1), 59–82.

Borovnik, M. (2022). Seafarers: The force that moves the global economy. In K. Peters, J. Anderson, A. Davies, & P. E. Steinberg, *The Routledge handbook of ocean space* (pp. 148–160). Routledge.

Borovnik, M. (2024). Trapped in the COVID-19 pandemic: Seafarers and the global crew change crisis. *Asia Pacific Migration Journal, 33*(1) 93–117.

Borovnik, M., Bedford, C., & Bailey, R. (2021). Has COVID-19 ended seafaring for Kiribati? *DEVPOLICY.* Retrieved May 23, 2024, from https://devpolicy.org/has-covid-19-ended-seafaring-for-kiribati-20211222/

Brooks, S. K., & Greenberg, N. (2022). Mental health and wellbeing of seafaring personnel during COVID-19: Scoping review. *Journal of Occupational Health, 64*(1), e12361. https://doi.org/10.1002/1348-9585.12361

Central Intelligence Agency (CIA). (2023). *The World Factbook: Kiribati*. Retrieved March 13, 2023, from https://www.cia.gov/the-world-factbook/countries/kiribati/#people-and-society

Change.Org. (2021). *Seafarers are essential* [video published 8 April 2021]. Retrieved May 23, 2024, from https://vimeo.com/533914260

Cook, N., & Butz, D. (2018). Moving toward mobility justice. In N. Cook & D. Butz (Eds.), *Mobilities, mobility justice and social justice* (pp. 3–21). Routledge.

Cresswell, T. (2001). *The tramp in America.* Reaktion Books.

Cresswell, T. (2016). The vagrant/vagabond: The curious career of a mobile subject. In T. Cresswell & P. Merriman (Eds.), *Geographies of mobilities: Practices, spaces, subjects* (pp. 239–253). Routledge.

Cresswell, T. (2010). Towards a politics of mobility. *Environment and Planning D: Society and Space, 28*(1), 17–31.

De Beukelaer, C. (2021). COVID-19 border closures cause humanitarian crew change crisis at sea. *Marine Policy, 132*, 104661. https://doi.org/10.1016/j.marpol.2021.104661

Devereux, H., & Wadsworth, E. (2022). Forgotten keyworkers: The experiences of British seafarers during the COVID-19 pandemic. *The Economic and Labour Relations Review, 33*(2), 272–289.
Doumbia-Henry, C. (2020). Shipping and COVID-19: Protecting seafarers as frontline workers. *Journal of Maritime Affairs, 19*, 279–293.
Foley, E. (2022, April 25). War in Ukraine: Seafarers suffering human rights violations need our continued support. *Global Maritime Forum.* Retrieved May 23, 2024, from https://www.globalmaritimeforum.org/news/war-in-ukraine-seafarers-suffering-human-rights-violations-need-our-continued-support
Gallez, C. (2023). Mobility justice as a political object. In G. Drevon & V. Kaufmann (Eds.), *Mobility and geographical scales* (pp. 37–61). ISTE Ltd.
Global Maritime Forum. (2023). *Neptun declaration.* Retrieved May 23, 2024, from https://www.globalmaritimeforum.org/neptune-declaration/crew-change-indicator
Graham, C. (2022). *A permanent fix for seafarers as essential workers.* Retrieved May 23, 2024, from https://splash247.com/a-permanent-fix-for-seafarers-as-essential-workers/
Harada, T. (2023). Mobility justice and sustainable futures. *Australian Geographer, 54*(4), 425–431.
Harvey, D. (2015). *Seventeen contradictions and the end of capitalism.* Oxford University Press.
Human Rights at Sea. (2022). *Ukraine: Blue safe corridors at sea.* Human Rights at Sea. Retrieved May 23, 2024, from https://www.humanrightsatsea.org/news/ukraine-blue-safe-corridors-sea
International Chamber of Shipping (ICS). (2022). *Coronavirus (COVID-19): Seafarer shore leave principles.* Retrieved May 23, 2024, from https://www.ics-shipping.org/wp-content/uploads/2021/09/Coronavirus-COVID-19-Seafarer-Shore-Leave-Principles.pdf
International Chamber of Shipping (ICS). (2021). *Shipping and world trade: Global supply and demand for seafarers.* Retrieved May 23, 2024, from https://www.ics-shipping.org/
International Labour Organization (ILO). (2023). *Ensure decent work for key workers.* Retrieved May 23, 2024, from https://www.ilo.org/global/about-the-ilo/newsroom/news/WCMS_871619/lang--en/index.htm
International Labour Organization (ILO), International Maritime Organization (IMO), United Nations Trade and Development (UNCTAD), & World Health Organization (WHO). (2022). *Joint Statement urging continued collaboration to address the crew change crisis, safeguard seafarer*

health and safety, and avoid supply chain disruptions, during the ongoing COVID-19 pandemic. Retrieved May 23, 2024, from https://unctad.org/system/files/non-official-document/un-joint-statement-on-crewing-crisis_en.pdf
International Maritime Organisation (IMO). (2021a). The World Maritime theme for 2021 is dedicated to seafarers, highlighting their central role in the future of shipping. *International Maritime Organisation News*. Retrieved May 23, 2024, from https://www.imo.org/en/MediaCentre/PressBriefings/pages/IMO-launches-a-year-of-action-for-seafarers.aspx
International Maritime Organisation (IMO). (2021b). *Seafarer access to medical care a matter of life and death*. Retrieved May 23, 2024, from https://www.imo.org/en/MediaCentre/PressBriefings/pages/medicalassistance.aspx
International Maritime Organisation (IMO). (2021c). *Pacific hub ports concept explored for regional crew change and repatriation*. Retrieved May 23, 2024, from https://www.imo.org/en/MediaCentre/Pages/WhatsNew-1587.aspx
International Maritime Organisation (IMO). (2021d). *400,00 seafarers stuck at sea as crew change crisis deepens*. Retrieved May 23, 2024, from https://www.imo.org/en/MediaCentre/PressBriefings/Pages/32-crew-change-UNGA.aspx
International Maritime Organisation (IMO). (2022). Learning the lessons from the COVID-19 pandemic. Online Access: https://www.imo.org/en/MediaCentre/PressBriefings/pages/Learning-the-lessons-from-the-COVID-19-pandemic.aspx
Jonglertmontree, W., Kaewboonchoo, O., Morioka, I., & Boonyamalik P. (2023). Depressive symptoms among Thai male seafarers during the COVID-19 pandemic: A cross sectional study. *BMC Public Health, 23*(475). https://doi.org/10.1186/s12889-023-15305-7.
Kitada, M. (2021). Women seafarers: An analysis of barriers to their employment. In V. O. Gekara & H. Sampson (Eds.), *The world of the seafarer: Qualitative accounts of working in the global shipping industry* (pp. 65–78). Springer.
Kotef, H. (2015). *Movement and the ordering of freedom: On liberal governances of mobility*. Duke University Press.
Lucas, D., Jego, C., Jensen, O. C., Loddé, B., Pougnet, R., Dewitte, J.-D., Sauvage, T., & Jegaden, D. (2021). Seafarers' mental health in the COVID-19 era: Lost at sea? *International Maritime Health, 72*(2), 138–141.
Luchenko, D., I., & Georgiievskyi. (2021). Administrative restrictions in ports: Practice of crew rotations during COVID-19 pandemic. *Lex Portus, 7*(3), 7–31.

Markkula, J. (2021). 'We move the world': The mobile labor of Filipino seafarers. *Mobilities, 16*(2), 164–177.
Ministry of Business, Innovation & Employment (MBIE). (n. d.). *MIQ timeline.* Retrieved May 23, 2024, from https://www.mbie.govt.nz/immigration-and-tourism/isolation-and-quarantine/managed-isolation-and-quarantine/about-miq/miq-timeline/
Munoz, T. (2022, October 10). Seafarers in crisis. *The maritime executive.* Retrieved May 23, 2024, from https://maritime-executive.com/magazine/seafarers-in-crisis
NZSeafarers. (2021). NZSeafarers on Facebook. Retrieved May 23, 2021, from https://www.facebook.com/NZSeafarers/.
Ogborn, M. (2008). *Global lives: Britain and the world, 1550–1800.* Oxford University Press.
Radio New Zealand (RNZ). (2021, April 12). *COVID-19 mariners organisation wants jabs* [audio]. Retrieved May 23, 2024, from https://podcast.radionz.co.nz/mnr/mnr-20210412-0652-covid-19_mariners_organisation_wants_jabs-128.mp3
Safety4sea. (2022). *Year in review: How the Russian invasion of Ukraine affected shipping in 2022.* Retrieved May 23, 2024, from https://safety4sea.com/cm-year-in-review-how-the-russian-invasion-of-ukraine-affected-shipping-in-2022/
Sampson, H. (2021). The rhythms of shipboard life: Work, hierarchy, occupational culture and multinational crews. In V. O. Gekara & H. Sampson (Eds.), *The world of the seafarer: Qualitative accounts of working in the global shipping industry* (pp. 87–98). Springer.
Sampson, H. (2024). *Sea-Time.* Routledge.
Schubert, K. (2021). Freedom as critique: Foucault beyond anarchism. *Philosophy and Social Criticism, 47*(5), 634–659.
Schuler, M. (2023). *China's rising COVID-19 infections causing concern for seafarer crew changes.* Retrieved April 7, 2024, from https://gcaptain.com/chinas-rising-covid-19-infections-causing-concern-for-seafarer-crew-changes/
SeafarerHelp. (2020). *Standing up for stranded seafarers – Human Rights Day 2020* [video]. Facebook. Retrieved May 23, 2024, from https://www.facebook.com/SeafarerHelp/videos/1299290523769513 /
Sheller, M. (2018a). *Mobility justice. The politics of movement in an age of extremes.* Verso.
Sheller, M. (2018b). Theorizing mobility justice. In N. Cook & D. Butz (Eds.), *Mobilities, mobility justice and social justice* (pp. 22–36). Routledge.

Sheller, M. (2023). Mobility justice after climate coloniality: Mobile communing as a relational ethics of care. *Australian Geographer, 54*(4), 433–447.
Shipping Australia. (2021). *Crew change crisis continues to worsen.* Retrieved May 23, 2024, from https://www.shippingaustralia.com.au/crew-change-crisis-continues-to-worsen/
Slišković, A. (2020). Seafarers' well-being in the context of the COVID-19 pandemic: A qualitative study. *Work, 67*(4), 799–809.
Soja, E. W. (2010). *Seeking spatial justice.* University of Minnesota Press.
Steinberg, P. E. (2015). Maritime cargomobilities: The impossibilities of representation. In T. Birtchnell, S. Savitzky, & J. Urry (Eds.), *Cargomobilities: Moving materials in a global age* (pp. 35–47). Routledge.
The World Bank. (2023). *Personal remittances, received (% of GDP) – Kiribati, 1979 – 2022.* Retrieved April 7, 2024, from https://www.cia.gov/the-world-factbook/countries/kiribati/#people-and-society
Torpey, J. C. (2000). *The invention of the passport: Surveillance, citizenship, and the state.* Cambridge University Press.
United Nations (UN). (n. d.). *Universal Declaration of Human Rights.* Retrieved May 23, 2024, from https://www.un.org/en/about-us/universal-declaration-of-human-rights
United Nations Conference on Trade & Development (UNCTAD). (2020). *Review of maritime transport 2020.* Retrieved May 23, 2024, from https://unctad.org/webflyer/review-maritime-transport-2020
United Nations Conference on Trade & Development (UNCTAD). (2022). *Review of maritime transport 2022: Navigating stormy waters.* Retrieved May 23, 2024, from https://unctad.org/system/files/official-document/rmt2022_en.pdf
United Nations Regional Centre for Preventive Diplomacy for Central Asia (UNRCCA). (2020). *Secretary-general's message on World Maritime Day, 24 September 2020.* Retrieved May 23, 2024, from https://unctad.org/webflyer/review-maritime-transport-2020
Urbina, I. (2019). *The outlaw ocean: Crime and survival in the last untamed frontier.* Random House.
Urry, J. (2008). Moving on the mobility turn. In W. Canzler, V. Kaufmann, & S. Kesselring (Eds.), *Tracing mobilities: Towards a cosmopolitan perspective.* Routledge.
Verlinghieri, E., & Schwanen, T. (2020). Transport and mobility justice: Evolving discussions. *Journal of Transport Geography, 87*, 102798. https://doi.org/10.1016/j.jtrangeo.2020.102798

Wasuka E. (2021). *Kiribati seafarers stranded in Australia.* ABC Pacific. Retrieved May 23, 2024, from https://www.abc.net.au/pacific/programs/pacificbeat/kiribati-aus-sailors/13647674

Wemyss G. (2023). Bordering seafarers at sea and onshore. *Frontiers in Sociology, 7*. https://doi.org/10.3389/fsoc.2022.1084598

Williams, A., & Nicholls, R. (2021). *Human rights at sea. Addressing the future of seafaring in Kiribati.* Independent Case Review by Human Rights at Sea. Retrieved May 23, 2024, from https://www.humanrightsatsea.org/sites/default/files/media-files/2021-12/HRAS_Kiribati_Seafarer_Futures_Case_Study_September_2021-SP-LOCKED.pdf

World Health Organization (WHO). (2023). *Timeline: WHO's COVID-19 response.* Retrieved May 23, 2024, from https://www.who.int/emergencies/diseases/novel-coronavirus-2019/interactive-timeline

Infrastructural Containment and the Politics of Migration in the Mediterranean Sea

Lara Şarlak

Introduction

In February 2020, the Greek government announced plans to build a 2.7-kilometre floating wall in the Aegean Sea to prevent migrants from reaching its islands via Turkey. After the announcement, human rights activists criticised Greece for what they described as an antimigrant policy. The wall, according to the widespread critiques, would create more dangers by halting rescue operations and life-saving assistance on an already-deadly terrain (Becatoros, 2020). Reporters have compared the floating barrier—which was projected to cost at least half a million euros to build and maintain—to Trump's infamous planned wall across the US-Mexico border (Hickman, 2020). While there are undeniable resemblances between the American and Greek plans, it is difficult

L. Şarlak (✉)
Department of Anthropology, The University of British Columbia, Vancouver, BC, Canada
e-mail: lara.sarlak@alumni.ubc.ca

K. Peters and J. Turner (eds.), *Ocean Governance (Beyond) Borders*,
https://doi.org/10.1007/978-3-031-71322-4_11

to ignore one major difference: the terrain on which these walls were planned to be built (Elden, 2021).

The materiality of the ocean demands unconventional modes of planning, engineering and maintenance: especially as the ideas that require planning, engineering and maintenance derive, oftentimes, from more conventional grounded, landed approaches. A fence, for example, is a typical technology of bordering, somewhat easily enacted on a stable surface like land. Yet, having a fence on the ocean's surface necessitates submerging pylons into the ocean floor to establish a metre-high barrier of nets with flashing lights to demarcate the floating wall that would otherwise be invisible at night (Block, 2020). In fact, Greece's plan to build a floating fence demonstrates the extent to which states develop ways to adapt to the spatial and material obstacles posed by the ocean itself to strengthen the borders of their territorial sovereignty at sea.

This chapter questions the role of oceanic terrain as an *environmental infrastructure*, wherein the natural environment becomes an integral component of the material infrastructure. Through a closer look at the borders in infrastructural terms, I explore how water's material distinctiveness impacts oceanic governance and strategies of containment aimed at preventing migrant mobilities, often developed through planning and decision-making on land. Here, containment refers to measures of confinement aimed at regulating and disrupting mobilities within and across borderzones and asserting control over migrants' autonomous presence in particular spaces. I follow Tazzioli and Garelli's (2020) theorisation of 'containment beyond detention', which explores containment practices that extend beyond the premises of traditional detention sites and their call for further academic research on migration containment. This also builds from theories in carceral geography, where 'carceral conditions' need not be limited only to traditional prison sites, but also any sites where spatiality, detriment and intent coalesce (Moran et al., 2018; Turner & Peters, 2018). I hence expand my analysis beyond the conventional understanding of containment as *typically* 'carceral' and shift perspective to the sea to question how spatial confinement of migrants additionally emerges from practices of neglect in watery spaces.

I developed the concept *infrastructural containment* to address the spatio-material strategies deployed by nation-states to regulate, block and push back clandestine maritime crossings and further strengthen border infrastructures. I argue that such containment is enacted through the assemblages of diverse technopolitical actors, materials and systems, marking territories both through physical and abstract forms that constitute the nation-state borders. The ocean as a hostile terrain—an environmental infrastructure—also plays an active role in facilitating and/or inhibiting infrastructural containment by acting as a natural barrier or by resisting nation-state boundaries. The EU's attempts to limit migrant mobility thus involves transforming the Mediterranean Sea into a site of control, often, as I will show, via modes of (in)visibility of that ocean terrain.

To illustrate these points, I analyse various methods of (un)making of maritime borders at and in the Mediterranean Sea to demonstrate how border infrastructures take advantage of the terrain's material and ontological characteristics and how states configure new strategies to prevent mobility while migrants seek new and safer ways to breach them. Accordingly, I think *with* and *through* the ocean (Peters & Steinberg 2019) to grasp how border infrastructures are constructed on a fluid surface and consider the many challenges that arise during their governance and maintenance. European governance in the Mediterranean Sea creates one of the deadliest borders in the world (Stierl, 2018, 2021). Many describe this hostile terrain as a 'graveyard' on account of the thousands of migrants who flee war, violence, famine and poverty from countries like Syria, Libya, Afghanistan, Pakistan, Eritrea and Somalia who have drowned while attempting to reach Europe. According to UNHCR (2022), between 2014 and 2021, more than 24,400 migrants have lost their lives or went missing while trying to reach Europe by crossing the Mediterranean Sea. For many, reaching Europe through rough seas has become a more prevalent choice due to the inaccessibility of terrestrial routes. Migrants travelling on foot often encounter more physical barriers such as fences, walls and checkpoints, which prolong their journey and heighten the challenge of unauthorised crossings. Additionally, given the geographical layout of Europe, surrounded by seas and oceans, albeit perilous maritime crossings can sometimes be the only viable option for migrants coming from regions without shared land borders with the EU. This is not to say that land crossings are more predictable and therefore safer for migrants. As many studies have shown, clandestine crossings on land can also be tremendously deadly when, for instance, migrants try to reach the United States from Mexico by crossing the dangerous Sonora Desert (De Leon, 2015; Sundberg, 2011) or to enter Greece from Turkey by crossing the Evros River (Pavlidis & Karakasi, 2019). According to the latest figures, in 2023, the number of migrant arrivals to Europe by sea is nine times greater than that of by land (IOM, 2023). Thus, even though ocean presents itself as a 'hostile terrain' (De Leon, 2015), the seemingly uncontrollable (and thus relatively porous) borders across the Mediterranean Sea make it a dangerous but viable route for escape compared to Europe's land borders.

While the dichotomy between land and ocean has been reflected in immigration policies (e.g. the wet foot/dry foot policy of the United States[1]), in settler-colonial states such as Australia, the excision of mainland from the migration zone as a strategy to keep asylum seekers offshore also brings the fixity of land borders into question (Mountz, 2013; Perera, 2013). Whether on land or water, territorial borders conceived as infrastructures are emergent systems in constant states of transformation, intricately interwoven with their environment. Therefore, this chapter demonstrates how border securitisation practices themselves are contingent on the ontology and material characteristics of the terrain on which border infrastructures are developed. I follow Dijstelbloem's (2021: 25) framing that 'Europe's technopolitical border networks' constitute 'a kind of infrastructure' that highlights the intimate connectivity of diverse policies and practices and that functions to organise 'preventions, selections, and interventions' of noncitizens trying to enter the EU. According to Dijstelbloem, border infrastructures are characterised by expansive networks that are inherently linked to local situations. They are designed to organise the circulation of migrants and intentionally exclude those who cannot access legal entry to the EU. Border infrastructures, like many other forms of infrastructures, are both visible and invisible. As evinced by the externalisation of the

[1] In 1995, the so-called wet foot/dry foot policy was implemented in the United States for Cuban migrants, allowing those who reached US soil ('dry foot') to stay but returning those intercepted at sea ('wet foot'). The policy that was aimed at deterring migrant arrivals to the United States via sea was terminated in 2017 (Dickson, 2021).

EU's border control, border infrastructures also present themselves as moveable entities. Configuring Europe's borders as infrastructures invites closer attention to the affective forces of non-human bodies, including that of the *oceanic terrain*, that are assembled to conceive the state's territorial boundaries.

The concept of infrastructural containment, when applied to waterscapes, sheds light on the 'conditions under which the sea was made to kill' (Heller & Pezzani, 2014: 678). In what follows, first, building on the anthropological literature on infrastructures, I introduce the infrastructural lens with which this paper examines maritime borders and discusses how borders transcend their fixed, linear locations. I then suggest that an infrastructural lens allows a better understanding of the containment of mobility enacted through the spatio-material form of borders at sea. By analysing the violent border securitisation practices, I explore the (un)making of borders at the Mediterranean Sea, conceived as an environmental infrastructure, and how the demarcation of nation-state boundaries enacts infrastructural containment on a hostile terrain. Throughout this analysis, the ocean itself and its (in)visibility especially appear as a crucial actor that materialises or resists border infrastructures.

An Infrastructural Approach to Water

The question of how different seas exist is a complex one given the fact that 'all the seas of the world run into each other'; they are all 'fundamentally connected [and] "indivisible" in Victor Hugo's words' (Mack, 2011: 37). How, then, are these vast bodies of water divided by imaginary (yet simultaneously very 'real' and 'hard' national) borders that, in the context of migration, are so gruelling to cross? I argue that this inquiry can be enriched through an analysis of maritime borders from an infrastructural perspective that accounts for the dynamic, fluid and relational qualities that constitute borderzones. This also allows for an analysis that is not entirely human-centered. Indeed, *environmental infrastructures* provide a useful framework to grasp the divergent ways in which human and non-human 'things' (migrants, boats, waves, salt, wind, ports, rescue operations, detention centres and so on) are assembled to construct a border across the Mediterranean Sea. In her study on clandestine infrastructures in the US-Mexico borderlands, Muehlmann (2019: 49) suggests that '"nature" can become infrastructure when placed in the context of particular networks of movement and exchange'. Oceanic space, in this respect, can be understood as an environmental infrastructure that is reconfigured and territorialised as maritime borders.

Approaching border securitisation practices at the Mediterranean Sea through an infrastructural lens, I question how oceans as 'wet' spaces are shaped into particular terrains of governance, which in turn enact specific modes of border enforcement. In this respect, drawing on the concept of 'wet ontology' as explored by Steinberg & Peters (2015: 252) is helpful to question the emergence of wet politics and 'enables us to recognize that the form of water opens new territories of control and conflict'. Wet ontologies and such 'oceanic thinking' challenge ontological assumptions that are predominantly terrestrial, static and linear and invite scholars to 'rethink the borders that we apply to various materialities and their physical states' (Steinberg & Peters, 2015: 259). Following this perspective, I suggest an infrastructural approach to water that serves as a polyvalent 'sociomaterial interface' (Boyle & Shneiderman, 2020: 115) to capture how borderzones are established through the assemblages of diverse actors, materials and systems, marking territories. An infrastructural lens identifies 'new ways of marking limits' (Reeves, 2017: 728). Thus, I bring the concept of infrastructures firmly to that of oceanic governance.

The ocean as vast, deep, material terrains, and water as a vital matter have perhaps unexpectedly been a key part of critical theoretical studies, ranging from the exploration of agencies in the seas (Alaimo, 2016) to discussions of 'wet' (Steinberg & Peters, 2015) and 'more-than-wet' ontologies (Peters & Steinberg, 2019). These studies offer novel ways of understanding flows, connections and fluidities that underline the dynamic assemblages sustained by and through hydrospheres such as the sea. Informed by this critical literature, I question how oceanic spaces and their liquidity are divided by territorial boundaries in the Mediterranean Sea. While the ocean's hostile physical terrain constitutes an intimidating border infrastructure—one that serves the needs of the 'fortress-makers'—its intractability and vastness simultaneously disrupt these borders as states 'fail' to control 'unauthorised' human mobility.

As Appel et al. (2018: 25) claim, 'matter and nonhuman relations are not just inert substrate that yields to the dreams and desires of powerful (human) actors. Instead, infrastructure's materials are active participants in its form, and therefore, also its politics'. Conceptualising the oceanic space as an environmental infrastructure thus positions the ocean as a significant actor in the formation of maritime borders. Juxtaposing infrastructure literature with posthumanist theorisations of oceanic space also blurs the boundaries between the environment and infrastructures that capture the entanglements of human and nonhuman actors forming maritime borders (Nelson & Bigger, 2022). As Dijstelbloem (2021: 26) argues, 'borders can also intermingle with nature and render different terrains into borders or stand out materially from their surroundings'. Such border-environment convergence can be clearly observed when international rivers are conveniently converted into borders dividing neighbouring states. For instance, in seeking to understand the dialectical relationship between rivers and borders, Thomas (2021) reconceptualises international rivers as border infrastructures throughout their analysis of the Ganges River, which fundamentally shapes the Indo-Bangladeshi borders.

The aesthetics of infrastructures also matter. Larkin (2018: 176) states that 'in the study of infrastructures, form is both ubiquitously visible yet absent from analytical consideration'. While imagining infrastructure beyond their physical presence is already challenging, this task is even more complicating when thinking of border infrastructures that are conceived in more intangible and abstract forms as opposed to concrete walls and electrified fences. Due to its turbulent fluidity, the ocean not only bears borders but also 'exceeds' and 'leaks' (Peters & Steinberg, 2019) into Euro-Mediterranean borderzones.[2] This unstableness also problematises the linear and static perception of borders that dominates western notions of the demarcating of space and territory.

[2] I recognise that certain physical land borders, such as the US-Mexico border, also have excess and leakages.

Migration scholars have been theorising how nation-state boundaries extend beyond the location of borders themselves (Ellebrecht, 2020; Mountz, 2013) and consider borders as mobile entities (Dijstelbloem, 2021). Such accounts can also find meaning through an infrastructural lens, although, I argue, it is a challenging endeavour due to the risk of reducing borders into physical organisational structures. Nevertheless, there have been notable works that tackle this issue. For example, Boyle and Shneiderman's (2020: 115) idea of 'infrastructural effects' helps to grasp the relationship between infrastructures and borderlands from a multidimensional and open-ended approach that '[extends] beyond the spatial and temporal constraints of the physical structures themselves'. Similarly, Dalakoglou's (2017) study on the cross-border highway between Greece-Albania complicates the conventional understanding of the rigidness of borders by investigating the 'fluidity' of migrants' houses in Albania. Further, drawing from her ethnographic research in Kyrgyzstan, Reeves (2017) investigates how the construction of so-called 'independent roads' along Kyrgyzstan's porous land-border with Tajikistan becomes forces of 'infrastructural intervention' in a disputed border region. Here, roads serve as de facto state borders in the absence of juridical determination of international boundaries. As they circumvent and/or bypass contested landscapes, such infrastructural interventions solidify territorial limits, creating new forms of separation and entrapment for people living in the border region. These studies use infrastructure as a way to trace various seemingly disconnected spaces that, in fact, mark territorial boundaries. They also complicate the more conventional idea of infrastructures as purely physical entities. Drawing on this literature, I next discuss how oceanic terrain—considered as an environmental infrastructure—can decipher migrant experiences by actively shaping their lives through its wetness. The ocean's slippery presence and its 'excess' (Peters & Steinberg, 2019) help to show how borders also extend beyond their fixed, linear locations.

Infrastructural Containment at Maritime Borders

Due to the fluid, vast and volatile form of oceans, territorial boundaries are scattered and arguably more uncontrollable when thinking of the maritime borders that divide the Euro-Mediterranean region in comparison to terrestrial borders. Given the ostensibly more unconstrained route at sea, nation-states have thus worked on advancing new methods and technologies to overcome the uncertainty of oceanic void and build border infrastructures *across* and *with/in* the Mediterranean Sea. These strategies are entangled with the 'wetness' of water, manifesting themselves in multifarious types of infrastructural boundary enforcement—the floating barrier I discussed in the opening paragraph is just one of many tactics to mark state territories that are challenged by the ocean's unruliness. Although an abandoned plan, the literal floating barrier demonstrates how border infrastructures are constantly transformed due to the affective power of the terrain on which the nation-state territories are demarcated. As the ocean 'unmistakably [undergoes] continual reformation' (Steinberg & Peters, 2015: 248), so do the border infrastructures that parcel out the Mediterranean Sea to mark the EU territory.

In this final section, I consider the affective power of the ocean's wet state and examine what it means for bordering in the sea. I use the notion environmental infrastructures explained in the preceding sections as an analytical lens to attend strategies of containing migrant mobility that take advantage of the ocean's materiality, which renders boat migration as perilous journeys at sea. These processes that animate border infrastructures lead to what I call a wider form of *infrastructural containment*, that is, the heterogeneous assemblages of boundary enforcement strategies, which, in this case, target migrants to regulate and disrupt their mobility on the ocean and keep them *away* from the European space. Oceanic terrain and its distinct materiality appear to play an active role as such mechanisms of control are shaped in conjunction to its spatio-material presence as a liquid, slippery and opaque abyss.

In this way, the ocean, transformed into a natural boundary, assumes the role of preventing migrants to reach the shore in safety. Although the Mediterranean Sea essentially acts as a geographical boundary, separating

EU member states from their non-member neighbours, it is through the EU's militarised border securitisation practices aimed at disrupting migrant mobilities that the sea is configured into an environmental infrastructure that extends borders and contributes to the territorialisation of a dangerous terrain. As Chambers (2008: 27) who conducted his research in the Mediterranean borderzones describes, 'to be at sea is to be lost, and to be in such a state is to be vulnerable to encounters we do not necessarily control'.

The EU has taken advantage of this sense of misplacement that shapes one's journey at sea, which is constantly threatened by the ocean's uncontrollable nature. When considering sea-crossings, many migrants are inevitably subjected to endless risks stemming from this lack of control. The weather, size of the waves, durability of the ship, migrants' health conditions and amount of fuel are all among the shifting variables that make the possibility of crossing borders like a gamble. As the ocean is 'a swirling landscape of uncertainty' (Alaimo, 2016: 112), the Mediterranean Sea thus serves the EU by acting as a natural barrier that works to prevent mobility with its inherent danger.

Moreover, the ocean's opaqueness creates engagements with modes of visibility and invisibility via complex patterns that are reflected in the formation of maritime borders. As socio-material systems that have been famously theorised as 'invisible' until failing, or breaking down (Bowker, 1994; Star, 1999), infrastructures necessarily involve a politics of (in)visibility, and border infrastructures are no exception. Although the idea of an inherent invisibility of infrastructures has long been contested (See Larkin, 2013), revisiting this question is useful when examining the securitisation of maritime borders as they similarly 'display a particular interplay between visibility and invisibility' (Dijstelbloem, 2021: 40). According to Larkin (2018: 186), 'visibility and invisibility are not ontological properties of infrastructures; instead, visibility or invisibility are made to happen as part of technical, political, and representational processes'. Interrogating the processes by which (in)visibilities are produced at maritime border infrastructures illuminates the role the ocean's *wetness* plays in bordering the Euro-Mediterranean region. Such questioning challenges the concrete, linear and static characterisations of borders and allows us to grasp affective capacity of the water's fluid form in geopolitical processes of boundary enforcement.

The ocean's opacity, which complicates any idealisations of the terrain as an 'empty insubstantial space that one can seamlessly see through' (Steinberg, 2014: n.p.), has prompted efforts to gather information on mobility at sea and to make this voluminous space more transparent: to make it more visible (often to authorities). With the aim of increasing 'situational awareness' of the Mediterranean Sea (FRONTEX, 2022: n.p.), the EU has developed surveillance technologies to render the ocean into a site of control *despite* challenges posed by its vast and turbulent form. In fact, founded in 2013, the EU's European Border Surveillance System (EUROSUR) transformed the Mediterranean Sea into one of the most-closely-surveilled and hence visualised maritime spaces in the world. While the substantial scale and fluidity of the oceanic terrain allow migrant boats to escape the EU's watchful eye, intelligence-driven technologies such as satellites, radars and uncrewed aerial vehicles are assembled with patrol vessels to scan the ocean's surface and better rule the oceanic space (Jumbert, 2018; Topak, 2014).

For migrant boats intercepted at sea, the EU has also established brutal detention facilities and increased police presence in transit countries from which migrants escape to overcome challenges posed by governing the ocean and to make the state's presence more visible 'extra-territorially' (Dickson, 2021; Dijstelbloem, 2021; Ellebrecht, 2020). As 'the EU's "sea border" extends to any country with which it shares an ocean' (Ellebrecht, 2020: 141), deploying coastguards and military police in neighbouring non-member states has become a strategy that is central to the EU's border securitisation practices. Indeed, the EU's externalisation of border control aligns with Dijstelbloem's (2021) claim that border infrastructures are movable constructs.

Certainly, these strict restrictions to control the EU's external borders have resulted in riskier means of travel such as migration by boat through rough seas and numerous violations of human rights during practices of detention and deportation of migrants (Ellebrecht, 2020). Although these violent border securitisation practices—as justified by the EU—supposedly prevent migrant deaths at sea in addition to 'tackling irregular migration' (FRONTEX, 2022: n.p.), Jumbert's (2018) study depicts that the EU's surveillance measures that can potentially supply life-saving information are not effectively used for rescue operations. For instance, in the case of the 'Left-to-Die boat' in 2011, the Mediterranean Sea was one of the most highly surveilled areas of sea in the world due to NATO's military operations in Libya (Heller et al., 2012). Despite numerous signals of distress transmitting the boat's location and interactions with military vehicles, migrants drifted on the ocean for two weeks and were left to die. In this instance, there is wilful neglect, rendering the hypervisibility of the sea invisible to those with power, demonstrating the fluid workings of the infrastructural containment described earlier. Migrants are contained at sea, left to die at sea, through processes linked to the environmental terrain or the infrastructure of the sea, melded with associated containment infrastructures such as intense surveillance mechanisms or detention facilities outsourced to non-member states.

The deliberate indifference of the EU towards distressed migrants at sea was once again highlighted more recently by the Pylos shipwreck that claimed 500 lives in June 2023. This disaster, despite being labelled as the 'worst tragedy ever' in the Mediterranean Sea, garnered little attention particularly in comparison to the OceanGate rescue mission,[3] which mobilised multi-million-dollar search-and-rescue efforts to save millionaires and received tremendous media coverage (Mahdawi, 2023). The OceanGate mission underscored the presence of western nations' extensive resources and advanced technologies for conducting surveillance operations, not only on the ocean's surface but also in extreme underwater conditions with high pressure and limited visibility—all to rescue the ultra-rich travellers, while these very resources are consistently denied to distressed migrants. Indeed, satellite imagery, radio signals and sealed court documents, in addition to survivor testimonies alleging that both FRONTEX and the Greek coastguards spotted the overcrowded boat hours before it sank, all suggest that this tragedy was preventable.

While intense surveillance techniques such as cameras, drones, satellites and motion sensors emphasise the threatening presence of technological border infrastructures that constantly monitor the ocean's material

[3] On June 18, 2023, OceanGate's submersible, Titan, imploded while on an expedition to explore the Titanic wreck in the North Atlantic Ocean off Newfoundland, Canada.

surface to intercept migrant boats and deter unauthorised crossings, the simultaneous disappearance of state actors and entities when migrants call for help also enacts modes of containment that disrupt the mobility of noncitizens through neglect. This neglect—of leaving at sea—is a deployment of environmental infrastructure through border containment. As evinced in recurring instances of 'Left-to-Die boat', the EU's deliberate ignorance of boats in distress is central to its border securitisation practices that have increasingly become humanitarianised and militarised simultaneously.

Given the absence of state-run rescue missions operating at the Mediterranean Sea and with the ever-increasing death toll, civil actors have introduced several humanitarian-driven initiatives such as the Migrant Offshore Aid Station, Médecins Sans Frontières and Sea-Watch to run search-and-rescue missions and turn the sea into a less deadly space (Stierl, 2018). Although these NGOs have played a crucial role in contesting the European governance at sea, they have been facing a major crackdown due to the EU's criminalisation of their operations (Hardt & Mezzadra, 2020). The EU not only refuses to save migrant lives at sea, but also actively hinders these humanitarian missions by forcing them to wait for days to disembark their rescue ships. Indeed, the sudden disappearance of physical infrastructures such as safe ports for rescue vessels in the interest of protecting the EU borders demonstrates how 'the perceived in/visibility of infrastructure is also situated and contingent' (Colven, 2020: 6). By ignoring numerous calls for help and refusing to assign rescue vessels a safe port, the EU marks its territorial boundaries, this time, through disappearance of state actors who supposedly ensure the safety of individuals travelling on a highly dangerous terrain. Here, the EU's approaches adapt to the ocean's distinctive materiality—its own environmental infrastructure is used as a border—containing mobility by exposing migrants to suffering and death on a fluid and unstable terrain.

Conclusion

Borders enacted at sea challenge the idea of water worlds as an unmanageable 'void' (Steinberg, 2001: 14). Indeed, as noticeable vast bodies of saltwater, oceans are certainly not unmanageable; yet due to their volatile, dynamic, uncontrollable and liquid form, the ocean's hydrosphere can allude and resist the territorial boundaries configured by states (see also Acton et al., 2019). Indeed, Greece's floating wall project is designed as a solution to take more control of mobilities on/across/through this terrain. Aimed at preventing and containing migrant arrivals, the floating wall is yet another attempt to further concretise and militarise a border infrastructure (akin to those on land) and *in addition to* the terrain of the sea itself, which effectively discourages hundreds of migrants from crossing it in the first instance. Here the sea might be thought of as an environmental infrastructure of containment and control.

Upon receiving heavy criticism against the floating barrier due to its overt antimigrant purpose, the Greek government decided to abandon its plan to build a wall on the sea (Georgoula, 2022). However, the EU has continued to expand policies and practices that render the Mediterranean Sea a perilous terrain for migrants who intend to reach Europe. In this chapter, by analysing various bordering strategies that turn the ocean into a dangerous natural boundary, I have demonstrated the simultaneous materialisation and disruption of maritime borders shaped by the ocean's affective forces. This chapter has expanded the inquiry of 'how nonhuman nature might inflect the politics of boundary enforcement' (Sundberg, 2011: 323) by adopting an infrastructural approach to water. Treating the oceanic space as an environmental infrastructure offers renewed understandings of the restrictions on freedom of movement and to the work of bordering at sea. Owing to this polyvalent approach, I have examined how diverse state actors have benefited from the ocean's wetness that are entangled with border infrastructures.

And it is not just in the EU and the Mediterranean Sea where we see the ways in which the environmental infrastructures and infrastructures of containment described play out violently. Another distinct example of oceanic space being manipulated into a natural boundary in the context of migration is the case of the *Bibby Stockholm*. This is an engineless barge introduced by the UK's Home Office to detain asylum seekers who are supposedly free to get off, but are not allowed to leave the port. The vessel, which resembles a floating prison, had previously served as workforce accommodation for oil and gas companies and had been repurposed in the Netherlands to detain asylum seekers between 2005 and 2012 (Askew, 2023). It reveals a rather explicit example of infrastructural containment on water. By entrapping migrants in a floating vessel, just offshore, the nation-state still leverages the ocean's wetness and fluid surface to function as an additional obstacle to the fences and guards that surround the barge. This overcrowded migrant prison also replicates the conditions of strandedness in crammed boats at sea and thus serves as a stark reminder of the traumatic boat journeys migrants have endured to reach Europe. Indeed, in the Netherlands, *Bibby Stockholm* gained a reputation for providing poor quality healthcare and terrible services, leading to frequent instances of hunger strikes and riots amongst the migrant detainees (Askew, 2023). More recently, in the UK, the sudden evacuation of asylum seekers shortly after they embarked on the barge due to a Legionella outbreak in the water system and the subsequent installation of fire escapes prior to their return, demonstrates the absence of adequate safety measures while attempting to keep migrants out of sight (Ross, 2023). These 'setbacks' on basic infrastructure also portray how the concept of a floating prison is challenged by the very environment on which the prison was established and how the containment of migrants often displays landed approaches despite being conceived on water.

Consequently, the concept of *infrastructural containment* developed in this chapter articulates strategies used by nation-states to strengthen border infrastructures aimed at containing unauthorised maritime crossings (and in the case of the *Bibby Stockholm,* movement in general). As Perera (2013: 65) eloquently puts it, 'the oceans are annexed to the global borderlands. Water is unmade, configured as a space of stasis, suspension, confinement, capture and death'. To better grasp this territorialisation, I have interrogated the processes by which (in)visibilities are produced through maritime border infrastructures and how modes of containment are enacted through abandonment of distressed migrants at sea.

Aiming to police and deter illegalised migrants from crossing the Mediterranean Sea and to reduce the opacity of the oceanic terrain by transforming it into a site of control, the EU has assembled diverse surveillance and policing mechanisms through which infrastructural containment operates. Indeed, rather than preventing migrant lives lost at sea, the EU's techniques of policing, such as its decision to ignore numerous requests to assign a safe port for rescue ships, suggest that the EU's border infrastructures *contain* mobility by invisibilising the state when migrants in distress at sea are in urgent need of assistance. My chapter reveals that 'invisibility and visibility are thus socially constructed and materially produced by state among other actors' (Colven, 2020: 6). The dangerous oceanic terrain is configured as a substantial part of maritime borders as it is given the role to stop boat migrations by '[exposing] noncitizens to a state-crafted geopolitical terrain designed to deter their movement through suffering and death' (De Leon, 2015: 28). I have thus demonstrated how European nation-states develop border securitisation strategies that are fundamentally contingent on the very terrain on which borders are established.

Acknowledgements This chapter is based upon research conducted as part of my doctoral project, which was carried out with the aid of a grant from the International Development Research Centre (IDRC), Ottawa, Canada.

Disclaimers The views expressed herein do not necessarily represent those of IDRC or its Board of Governors.

References

Acton, L., Campbell, L. M., Cleary, J., Gray, N. J., & Halpin, P. N. (2019). What is the Sargasso Sea? The problem of fixing space in a fluid ocean. *Political Geography, 68*, 86–100.

Alaimo, S. (2016). *Exposed: Environmental politics and pleasures in posthuman times*. University of Minnesota Press.

Appel, H., Anand, N., & Gupta, A. (2018). Introduction: Temporality, politics, and the promise of infrastructure. In N. Anand, A. Gupta, & H. Appel (Eds.), *The promise of infrastructure* (pp. 1–38). Duke University Press.

Askew, J. (2023, August 7). The chequered past of Europe's 'floating prison'. *euronews*. Retrieved January 9, 2024, from https://www.euronews.com/2023/08/07/bibby-stockholm-the-chequered-past-of-europes-floating-prison

Becatoros, E. (2020, January 30). Rights group criticizes Greek sea barrier plan for migrants. *AP NEWS*. Retrieved January 9, 2024, from https://apnews.com/article/cf1d76ef8f6dbe892bb482c930dc2d11

Block, I. (2020, February 10). Greece plans floating sea border wall to keep out refugees. *Dezeen*. Retrieved January 9, 2024, from https://www.dezeen.com/2020/02/10/greece-floating-sea-border-wall-news/

Bowker, G. C. (1994). *Science on the Run: Information management and industrial geophysics at Schlumberger, 1920–1940*. The MIT Press.

Boyle, E., & Shneiderman, S. (2020). Redundancy, resilience, repair: Infrastructural effects in borderland spaces. *Verge: Studies in Global Asias, 6*(2), 112–138.

Chambers, I. (2008). *Mediterranean crossings: The politics of an interrupted modernity*. Duke University Press.

Colven, E. (2020). Subterranean infrastructures in a sinking city: The politics of visibility in Jakarta. *Critical Asian Studies, 52*(3), 311–331.

Dalakoglou, D. (2017). *The road: An ethnography of (im)mobility, space, and cross-border infrastructures in the Balkans*. Manchester University Press.

De Leon, J. (2015). *The land of open graves: Living and dying on the migrant trail.* University of California Press.

Dickson, A. J. (2021). The carceral wet: Hollowing out rights for migrants in maritime geographies. *Political Geography, 90*, 102475. https://doi.org/10.1016/j.polgeo.2021.102475

Dijstelbloem, H. (2021). *Borders as infrastructure: The technopolitics of border control*. The MIT Press.

Elden, S. (2021). Terrain, politics, history. *Dialogues in Human Geography, 11*(2), 170–189.

Ellebrecht, S. (2020). *Mediated bordering: Eurosur, the refugee boat, and the construction of an external EU border*. Transcript.

FRONTEX. (2022). *Monitoring and risk analysis*. FRONTEX European Border and Coast Guard Agency. Retrieved January 9, 2024, from https://www.frontex.europa.eu/what-we-do/monitoring-and-risk-analysis/monitoring-and-risk-analysis/

Georgoula, D. F. (2022). Building walls at sea: An assessment of the legality of the Greek floating barrier. *International Journal of Refugee Law, 34*(1), 54–81.

Hardt, M., & Mezzadra, S. (2020). Introduction. *South Atlantic Quarterly, 119*(1), 168–175.

Heller, C., & Pezzani, L. (2014). Liquid traces: Investigating the deaths of migrants at the EU's maritime frontier. In Forensic Architecture, A. Franke, E. Weizman, & Haus der Kulturen der Welt (Eds.), *Forensis: The architecture of public truth* (pp. 657–684). Sternberg Press.

Heller, C., Pezzani, L., & Situ Studio. (2012, April 4). The Left-to-Die boat. *Forensic Architecture*. Retrieved January 9, 2024, from https://forensic-architecture.org/investigation/the-left-to-die-boat

Hickman, M. (2020, February 11). Greece floats plan for refugee-deterring sea wall. *Archpaper.com*. Retrieved January 9, 2024, from https://archpaper.com/2020/02/greece-floats-plan-for-refugee-deterring-sea-wall/

IOM. (2023). *Migration flow to Europe: Arrivals*. Displacement Tracking Matrix. Retrieved January 9, 2024, from https://dtm.iom.int/europe/arrivals?type=arrivals

Jumbert, M. G. (2018). Control or rescue at sea? Aims and limits of border surveillance technologies in the Mediterranean Sea. *Disasters, 42*(4), 674–696.

Larkin, B. (2013). The politics and poetics of infrastructure. *Annual Review of Anthropology, 42*(1), 327–343.

Larkin, B. (2018). Promising forms: The political aesthetics of infrastructure. In N. Anand, A. Gupta, & H. Appel (Eds.), *The promise of infrastructure* (pp. 175–202). Duke University Press.

Mack, J. (2011). *The sea: A cultural history*. Reaktion Books.

Mahdawi, A. (2023, June 22). The Greek shipwreck was a horrific tragedy. Yet it didn't get the attention of the Titanic story. *The Guardian*. Retrieved January 9, 2024, from https://www.theguardian.com/commentisfree/2023/jun/22/the-greek-shipwreck-was-a-horrific-tragedy-yet-it-didnt-get-the-attention-of-the-titanic-story

Moran, D., Turner, J., & Schliehe, A. K. (2018). Conceptualizing the carceral in carceral geography. *Progress in Human Geography, 42*(5), 666–686.

Mountz, A. (2013). Shrinking spaces of asylum: Vanishing points where geography is used to inhibit and undermine access to asylum. *Australian Journal of Human Rights, 19*(3), 29–50.

Muehlmann, S. (2019). Clandestine infrastructures: Illicit connectivities in the US-Mexico borderlands. In K. Hetherington (Ed.), *Infrastructure, Environment, and Life in the Anthropocene* (pp. 45–65). Duke University Press.

Nelson, S. H., & Bigger, P. (2022). Infrastructural nature. *Progress in Human Geography, 46*(1), 86–107.

Pavlidis, P., & Karakasi, M.-V. (2019). Greek land borders and migration fatalities – Humanitarian disaster described from the standpoint of Evros. *Forensic Science International, 302*, 109875. https://doi.org/10.1016/j.forsciint.2019.109875

Perera, S. (2013). Oceanic Corpo-graphies, Refugee bodies and the making and unmaking of waters. *Feminist Review, 103*(1), 58–79.

Peters, K., & Steinberg, P. E. (2019). The ocean in excess: Towards a more-than-wet ontology. *Dialogues in Human Geography, 9*(3), 293–307.

Reeves, M. (2017). Infrastructural hope: Anticipating 'independent roads' and territorial integrity in Southern Kyrgyzstan. *Ethnos, 82*(4), 711–737.

Ross, A. (2023, September 13). Bibby Stockholm barge made ready to rehouse asylum seekers after legionella outbreak. *The Independent*. Retrieved January 9, 2024, from https://www.independent.co.uk/news/uk/home-news/bibby-stockholm-asylum-seekers-barge-b2410796.html

Star, S. L. (1999). The ethnography of infrastructure. *American Behavioral Scientist, 43*(3), 377–391.

Steinberg, P. E. (2001). *The social construction of the ocean.* Cambridge University Press.

Steinberg, P. E. (2014, October 18). *Volume, depth, and the (in)visibilities of water.* Retrieved January 9, 2024, from https://philsteinberg.wordpress.com/2014/10/18/volume-depth-and-the-invisibilities-of-water/

Steinberg, P. E., & Peters, K. (2015). Wet ontologies, fluid spaces: Giving depth to volume through oceanic thinking. *Environment and Planning D: Society and Space, 33*(2), 247–264.

Stierl, M. (2018). A fleet of Mediterranean border humanitarians. *Antipode, 50*(3), 704–724.

Stierl, M. (2021). The Mediterranean as a carceral seascape. *Political Geography, 88*, 102417. https://doi.org/10.1016/j.polgeo.2021.102417

Sundberg, J. (2011). Diabolic *Caminos* in the desert and cat fights on the Río: A posthumanist political ecology of boundary enforcement in the United States-Mexico Borderlands. *Annals of the Association of American Geographers, 101*(2), 318–336.

Tazzioli, M., & Garelli, G. (2020). Containment beyond detention: The hotspot system and disrupted migration movements across Europe. *Environment and Planning D: Society and Space, 38*(6), 1009–1027.

Thomas, K. A. (2021). International rivers as border infrastructures: En/forcing borders in South Asia. *Political Geography, 89*, 102448. https://doi.org/10.1016/j.polgeo.2021.102448

Topak, Ö. E. (2014). The Biopolitical border in practice: Surveillance and death at the greece-turkey borderzones. *Environment and Planning D: Society and Space, 32*(5), 815–833.

Turner, J., & Peters, K. (Eds.). (2018). *Carceral mobilities: Interrogating movement in incarceration.* Routledge.

United Nations High Commissioner for Refugees (UNHCR). (2022). *No end in sight.* Retrieved January 9, 2024, from https://storymaps.arcgis.com/stories/07502a24ce0646bb9703ce96630b15fa

Conclusion: Openings—Ocean Governance (Beyond) Borders

Kimberley Peters and Jennifer Turner

Afterthoughts

As editors of this collection, having worked collaboratively on several projects (see Peters & Turner, 2015; Turner & Peters, 2016, 2017 as examples), in 2020 we had the opportunity to teach together at the Carl von Ossietzky Universität Oldenburg, Germany. We developed a module titled 'Ocean Governance and Policy' where students—primarily from the natural sciences—would explore the social and political dimensions of the marine worlds that were a large part of their studies. The module was designed to help students 'understand what governance and policy are and what can be achieved through governance and the creation of

K. Peters (✉)
Marine Governance, Helmholtz Institute for Functional Marine Biodiversity at the University of Oldenburg, Oldenburg, Germany
e-mail: kimberley.peters@hifmb.de

J. Turner
Cultural and Political Geography, Universität Trier, Trier, Germany
e-mail: turner@uni-trier.de

K. Peters and J. Turner (eds.), *Ocean Governance (Beyond) Borders*,
https://doi.org/10.1007/978-3-031-71322-4_12

policy' (Ocean Governance and Policy Handbook, 2021: 1). Notably the module aimed to fulfil this goal through a deep, critical consideration of the topic, assisting students to understand,

> ...the logics and power dynamics that shape governance and policy, the politics of data that inform it, the difficulties of enacting governance and policy in a space that is (mostly) liquid, three-dimensional in form and has variable legal status, and where enforcement of governance is tricky in a space typically 'out of sight and mind'... It dismantles the idea of governance and policy as natural 'givens' to saving and fixing the oceans, to better interrogate how this goal can actually be achieved. The course aims to enable students to have a good handle on key issues and to be armed with knowledge to assist in more careful planning of marine futures (Ocean Governance and Policy Handbook, 2021: 1).

This module continues to be taught now, both in Oldenburg and also as a new course at the Universität Trier, Germany, called 'Human Geographies of the Ocean'. As part of both modules, students explore the dynamics of ocean bordering in a set of lectures entitled 'Spatial logics: Geography, governance and regimes of enclosure and freedom' and 'Landed critiques: Time, space and movement in ocean governance and policy'. At the end of these sessions, students are asked to engage in a debate. Framed within the United Nations Decade for Ocean Science and the global call to protect 30% of the oceans by 2030 (see Page, 2023), the debate splits the class into groups and asks each to take a different position point in relation to the following governance topic: *The desire to protect up to 30% of the oceans by the end of the decade should be achieved with static, fixed management measures (clearly demarcated plans and zones) rather than more dynamic, mobile, flexible approaches.* Each group hence presents either *for* or *against* the statement above (i.e. for static, fixed management, or against, and in favour of dynamic approaches).

What usually results from this task—now implemented in the last five academic years—is a shift in student perspective. Typically, shaped by what they read in the press of dominant conservation discourses, students believe bordered conservation measures to be the most suitable to 'save' and 'fix' imperilled seas. However, by unpacking such regimes critically, opinions start to change. Encouraged to think beyond their usual *limits*, students start to ask of the marine realm: 'who produces knowledge about the space, who controls it, how, and for whose benefit?' (Gray, 2018: 269). They consider that '[i]f imagining the high seas for conservation means demarcating and appropriating space, and claiming it for conservation' then questions of intent should be asked (ibid., 2018: 269). Students begin to see the downsides of bordering: that borders may not create 'real' zones of marine protection, but rather be paper parks on the journey to meeting global targets (for example see di Cintio et al. 2023; Mason, 2019; Matz-Lück & Fuchs, 2014). They acknowledge that conservation areas may be ineffective in view of migrating species (Crespo et al., 2020; Maxwell et al. 2015; Pendoley et al., 2014), and where species may move under the conditions of climate change (De Santo, 2012). They problematise how MPAs may be used geopolitically (De Santo, 2020; Leenhardt et al., 2013). They recognise that offshore spaces are hard to monitor and assess in terms of whether protection zones 'work' (Hilborn, 2018). They learn that MPAs may not meet the needs of, and may cause harm to, Indigenous, traditional, local, artisanal and small-scale fishers (Satizábal, 2018). Put simply: they see borders in their complexity and start to think beyond them. This is not to say the debate on MPAs is is straightforward (or that the module presents a view that is 'anti' MPA, see also De Santo 2024 on such discussions). Rather, students are encouraged to think more critically about MPA bordering operations.

The authors in this collection have explored enclosures across ten chapters to *open up* questions of how borders work, and to ruminate on alternatives: to think both with and beyond borders. Yet, those alternatives may be hard to envision or make possible against the tide of further bordering. As McAteer notes in his chapter (see also Peters, 2021 and Satizábal & Peters 2022)—and as echoed here in our reflections of a teaching task—conservation measures rest on further bordering with plans to enclose 30% of the oceans by 2030. National borders appear to be tightening with a shift to the Right and against migration in many northern European nations. The seas are enrolled in these articulations. Borders, it seems then, are here to stay. The chapters that comprise this book are nevertheless important in that they show how, oftentimes, borders may make little sense and can cause harm (even when they may be designed for the contrary). Borders are unrelenting. As such, it is vital to examine and interrogate them and to do so critically.

The chapters—and their authors—have hence demonstrated the *(un)naturalness* of lines. First, they have showed their taken-for-granted dominance as a 'solution' to fixing (in place) ocean issues. And, simultaneously, they have illuminated their unsuitability sometimes, for a geophysically, motionful, three-dimensional, state-shifting space that is also deep in its temporal attributes, ebbing and flowing back and forth in time as waves of history and possible futures are enfolded within its materiality. In a wide-ranging set of contributions, the book has revealed that borders are emplaced in oceans, through regimes of power—the physical building of infrastructures (of nets and fences to regulations and policy documents), to the construction of ideas that underpin them. Authors in this volume have illustrated that borders—in their many guises—result in a variety of conflictual outcomes as insides and outsides are created, sustained, but also pushed back against: challenged and resisted, by human and more-than-human life.

Beyond the examples used in the book's introduction—of marine protected area (MPA) management and asylum controls—this text has, throughout the previous ten chapters, charted a course across the oceans, revealing and highlighting the *multitude* of spatial, temporal, socio-political, material and ideological lines that are etched in the oceans. The book has, of course, taken on these examples, showing the significant complexities in how oceans border (and are bordered) in efforts to control people, resources and space (see "Imaginaries: Oceanic Bordering with Large-Scale Marine Protected Areas", by Montana and Davies on MPAs, and "Infrastructural Containment and the Politics of Migration in the Mediterranean Sea", by Şarlak on migration governance). Indeed, in these chapters, authors have explained that MPAs and asylum regulations may be commonplace examples, but their articulations are far from straightforward. It is not, or ever, simply a case of ocean lines being drawn for conservation or migration management, rather there are dynamics, power-dimensions, and violences to the ways lines are used in these contexts (by and for nations) that should be interrogated to better understand the work of borders and their a/effects. For Montana and

Davies, reflecting on the place of the nation in ocean governance, the Large Scale Marine Protected Area (known as a LSMPA, a phenomenon of creating huge areas, often in excess of 100,000 km^2, [see Smyth & Hanich 2019: 5]) is more than a designated space for marine protection. The (national) imaginations that feed into MPAs are a 'tool of governance' themselves (see "Imaginaries: Oceanic Bordering with Large-Scale Marine Protected Areas"). Accordingly, as Montana and Davies write, 'there is more at stake in the establishment of LSMPAs than conservation and marine management alone. The oceanic bordering agendas associated with LSMPAs not only redraw ocean maps, but also reinforce and renew national identities in relation to ocean space'. For Şarlak, also reflecting on the role of nations in bordering, the ocean itself becomes a border, used as a mechanism of 'infrastructural containment' for halting migratory movements. It is harnessed by authorities as a border in itself. As Şarlak writes, 'oceanic materialities are part of the complexity of migration in/at the Mediterranean Sea'. The ocean itself is enrolled by governments as part of 'spatio-material strategies deployed by nation-states to further strengthen borders and contain "clandestine" maritime crossings' (see "Infrastructural Containment and the Politics of Migration in the Mediterranean Sea"). Borders, then, even in these most recognisable examples of ocean line drawing, should not be overlooked. It is not enough to know they are there. These two chapters alone indicate that the critical unpacking of borders reveals their work, and hence raises questions of their logics (see also Peters, 2020). Borders create ways of knowing and understanding, representing and practising: they establish ideas, impact people, shape places and write futures.

The chapters *beyond* these ones likewise unpick the complexities of bordering, revealing borders to be far from innocent. In "Bordered-In, Bordered-Out, and Overlapping Territorialities in Ocean Space: The Case of Fisheries", Hung elucidates fisheries management—perhaps another mainstream example of how borders are put to work as a mode of regulation—as more than a straightforward way of governing resource extraction towards 'sustainable' ends. Rather Hung shows that *how* management occurs depends on *where* borders are enacted at sea (within or beyond national jurisdiction), and by *who*, and in whose benefit. Undertaking a review of the recent, critical literature on fisheries,

Hung argues that bordering oceanic resources and their extraction is an act of producing territory, which sees 'mobile state governance' being played out in global oceans, creating uneven outcomes for ocean users and dwellers. And, using a diverse range of case studies and contexts, other chapters illustrate how ocean borders move beyond examples we may readily associate them with. As Hung and Lien have noted (2022: 870) 'scholarship has shown how the immense and seemingly boundless oceans are in actuality subject to a variety of bordering forces' (Hung & Lien, 2022: 870). For example, this variety is revealed by Kunz ("Can Borders in the Ocean Respond to Climate Change?") who, although also writing of MPAs, asks the provocative question of whether these clearly bounded marine protected areas may be used to tackle the more boundless effects of *climate change* at a moment when they are being touted as a possible solution to such a global challenge. How, Kunz asks, can scales of action be transcended through the creation of more localised or regional bordering devices—or can they not? Similarly, Ansong shifts to different 'ground' ("Contested Borders and Resolution in Planning Shared Marine Waters"), to trace examples of how adjacent nations plan their coastal marine waters, where those waters have shared uses and ecosystems. How does marine spatial planning (MSP)—one of the most prominent modes of organising marine space currently—deal with bordering? He shows the somewhat arbitrary nature of lines, and shows, surprisingly, that some border designations are outstanding—lines are left undefined—when no resolution can be found. This raises, of course, a question of whether lines can do the job they are supposed to, prompting the recognition that they may fall short in the aspirations that some (i.e. government officials, policymakers, marine protection NGOs etc.) may hold of them.

Although not explicitly asking if we might move beyond borders, Ansong's chapter calls them into question. So too does Borovnik's contribution ("Borders and Confinement in Seafarers' Realities"). Bringing justice to centre-stage, Borovnik presents the violent irony of borders when the world's workforce, responsible for moving '90% of everything' (George, 2013), were held offshore by borders—suspended in space and time—during the COVID-19 pandemic. Savitzky et al.'s contribution ("Bordering marine belonging: The meanings, mobilities and materialities of bioinvasion") also tackles bordering *against* threats: the desire to use borders to create insides and outsides, to protect given 'insides' from certain 'outsides'—defining life that is not deemed to belong (see also "Introduction: Closures—Ocean Governance Borders"). Taking on the under-examined topic of biosecurity measures in light of 'threats' of marine invasive species the authors consider the normative assumptions on which invasive species management rests. They illustrate how preventing the movement of microorganisms in ballast tanks also relies as much on ideological borders of what belongs where, as on any quantitative evidence of harm to 'invaded' ecosystems as movement happens alongside the churn of the global economy ("Bordering Marine Belonging: The Meanings, Mobilities and Materialities of Bioinvasion", Savitzky et al.). Likewise taking on a less obvious articulation of oceanic bordering, Verne ("Human-Shark Encounters Beyond Borders: (Post-humanist) Attempts to Navigate a Maritime Contact Zone") shows how shark management to 'protect' leisure and tourism pursuits also rests on notions of threat that should be contained (with the opening of the chapter challenging the very basis of such threats in the first place). Verne unpacks the place of technologies in managing human-shark encounters, but notably, she raises the question of whether there may be other ways to think through these human-more-than-human interactions.

It is this assertion—that there could be ways to govern *beyond* borders, or at least to problematise and question borders—that drove the desire to convene this book (and its associated conference sessions). At the outset of this collection, McAteer ("Overdetermined by Territory? Governing the Ocean in Time, Matter, and Rhythm") and Couling ("Counter-mapping: A Morphology of Oscillating Margins in the Norwegian Sea") show that, whilst having a long, entwined and complex history, oceans and borders also sit uncomfortably together. As McAteer notes, the temporalities and materialities of the ocean—which defy easy bordering—may push us to think of other ways of governing. Likewise, Couling already demonstrates such potential with maps that might be better suited to the ocean's form—its depth and volume, its gradients, salinity and temporal form. Here she posits an approach in which we might move away from the flat depictions of oceans that are oftentimes used when drawing the oceans and lines *upon* them, to deeper modes of representation. Together the chapters confirm that '[a]n attention to zones, areas, sectors, borders, boundaries and limits; their ontological assumptions and stabilities in the realm of ocean governance may allow us to push those limits and to imagine – and in turn build – different governance futures' (Peters, 2020: 8).

What those futures may look like, remains to be seen. Can there be a future without ocean borders? Although we cannot answer this, what is clear is the need to continue the longstanding efforts highlighted by Steinberg (since 1999) and more recently by Hung and Lien (2022) to take ocean borders seriously. As the latter write 'recent thinking about borders as both an analytic tool and theoretical approach has highlighted the importance of maritime borders' (Hung & Lien, 2022: 870). The examples featured in this book are not the only ones that can and should be the focus of attention. Other empirical areas of study that tackle cross-border relations could be examined—for example, the transgression of land-sea borders through eutrophication (run-off into marine systems) and (plastic) pollution and other modes of waste that enter water worlds from the land. These examples demonstrate how matter both moves through and across geophysical borders and can be hard to contain for management purposes offshore, when it may transcend the borders marking jurisdictions of responsibility. Likewise, there is a need to take more seriously vertical borders with the advent of sea level rise as borders of land are subsumed by sea. The undersea is also a space that requires interrogation of how borders may be put to task. Currently, the International Seabed Authority (ISA) designates areal zones of seafloor for exploration in view of potential exploitation, but mining operations never exist within neat parcels of flat space rather through the water column and where mining plumes (the debris of extraction) move beyond borders (see Saputra & Sammler, 2024). This book has also not tackled all the cases by which border surveillance works (such as maritime policing, for example, against drug smuggling, or illegal, unreported and unregulated (IUU) fishing) or examined the relationship between borders and how satellite surveillance technologies are put to work. There is also a need, as the introduction notes, to fully consider the ways that bordering rests on specific logics of western reasoning (Lambach, 2021) meaning that there are—of course—other ways to think of how the oceans may be lived with. These derive from those whose worlds are already oceanic and entangled with the seas (George & Wiebe, 2020; Hau'ofa 2008).

Indeed, as uses of the ocean continue to proliferate through human movement to resource extraction, and as borders remain dominant in their deployment, questions of bordering practice are vitally relevant. Studies are necessary not just for questioning whether borders 'work' and if we might govern 'better', but for highlighting whether there are ways to govern that are unlocked from colonial logics that have driven bordering regimes at sea in the first instance (George & Wiebe, 2020; Khalili, 2021; Lambach, 2021; Peters, 2020). Posing such questions may make it possible to see alternatives that enliven more just, fair and sustainable means of caring for our seas, those who cross them, and the lives lived with(in) them. This book has investigated how we are 'locked into' accepted modes of management, creating ever more enclosed or, even, 'carceral' seas. Yet, it also offers multiple visions of ocean governance for people and planet (and links between them) that quiz these logics and challenge them. *Ocean Governance (Beyond) Borders* hence advocates further work, marking a continued critical shift in thinking about what is possible in ocean governance—and the need, quite literally, to think outside of the box.

References

Crespo, G. O., Mossop, J., Dunn, D., Gjerde, K., Hazen, E., Reygondeau, G., Warner, R., Tittensor, D., & Halpin, P. (2020). Beyond static spatial management: Scientific and legal considerations for dynamic management in the high seas. *Marine Policy, 122*, 104102. https://doi.org/10.1016/j.marpol.2020.104102

De Santo, E. M. (2012). From paper parks to private conservation: The role of NGOs in adapting marine protected area strategies to climate change. *Journal of International Wildlife Law & Policy, 15*(1), 25–40.

De Santo, E. M. (2020). Militarized marine protected areas in overseas territories: Conserving biodiversity, geopolitical positioning, and securing resources in the 21st century. *Ocean & Coastal Management, 184*, 105006. https://doi.org/10.1016/j.ocecoaman.2019.105006

De Santo, E. M. (2024). *Securitizing Marine Protected Areas: Geopolitics, Environmental Justice, and Science*. Routledge.

Di Cintio, A., Niccolini, F., Scipioni, S., & Bulleri, F. (2023). Avoiding "paper parks": A global literature review on socioeconomic factors underpinning the effectiveness of marine protected areas. *Sustainability, 15*(5), 4464. https://doi.org/10.3390/su15054464

George, R. (2013). *Deep Sea and foreign going: Inside shipping, the invisible industry that brings you 90% of everything*. Portobello Books.

George, R. Y., & Wiebe, S. M. (2020). Fluid decolonial futures: Water as a life, ocean citizenship and seascape relationality. *New Political Science, 42*(4), 498–520.

Gray, N. J. (2018). Charted waters? Tracking the production of conservation territories on the high seas. *International Social Science Journal, 68*(229–230), 257–272.

Hau'ofa, E. (2008). *We are the ocean: Selected works*. University of Hawaii Press.

Hilborn, R. (2018). Are MPAs effective? *ICES Journal of Marine Science, 75*(3), 1160–1162.

Hung, P. Y., & Lien, Y. H. (2022). Maritime borders: A reconsideration of state power and territorialities over the ocean. *Progress in Human Geography, 46*(3), 870–889.

Khalili, L. (2021). *Sinews of war and trade: Shipping and capitalism in the Arabian Peninsula*. Verso Books.

Lambach, D. (2021). The functional territorialization of the high seas. *Marine Policy, 130*, 104579. https://doi.org/10.1016/j.marpol.2021.104579

Leenhardt, P., Cazalet, B., Salvat, B., Claudet, J., & Feral, F. (2013). The rise of large-scale marine protected areas: Conservation or geopolitics? *Ocean & Coastal Management, 85*, 112–118.

Mason, P. B. (2019). *UK marine protected areas: Powerful legal protection or paper parks?* (Doctoral dissertation). University of Essex. Retrieved June 23, 2024, from https://repository.essex.ac.uk/25323/1/Peter%20Mason%20PhD%20Law%20Thesis%20final%20submission.pdf

Matz-Lück, N., & Fuchs, J. (2014). The impact of OSPAR on protected area management beyond national jurisdiction: Effective regional cooperation or a network of paper parks? *Marine Policy, 49*, 155–166.

Maxwell, S. M., Hazen, E. L., Lewison, R. L., Dunn, D. C., Bailey, H., Bograd, S. J., & Benson, S. (2015). Dynamic ocean management: Defining and conceptualizing real-time management of the ocean. *Marine Policy, 58*, 42–50.

Ocean Governance and Policy Handbook. (2021). From the Marine Environmental Science Masters Programme, Carl von Ossietzky Universität Oldenburg.

Page, R. W. (2023). *30x30: From global ocean treaty to protection at sea.* Greenpeace. Retrieved 23 June, 2024, from https://www.greenpeace.org/international/publication/62121/30x30-from-global-ocean-treaty-to-protection-at-sea/

Pendoley, K. L., Schofield, G., Whittock, P. A., Ierodiaconou, D., & Hays, G. C. (2014). Protected species use of a coastal marine migratory corridor connecting marine protected areas. *Marine Biology, 161*, 1455–1466.

Peters, K. (2020). The territories of governance: Unpacking the ontologies and geophilosophies of fixed to flexible ocean management, and beyond. *Philosophical Transactions of the Royal Society B, 375*(1814), 20190458. https://doi.org/10.1098/rstb.2019.0458

Peters, K. (2021). A line in the ocean 30×30: Ocean borders and geography's limits. *Chair's plenary lecture at the Royal Geography Society (with IBG) Annual International Conference*, London, August 31–September 3, 2021.

Peters, K., & Turner, J. (2015). Between crime and colony: Interrogating (im)mobilities aboard the convict ship. *Social & Cultural Geography, 16*(7), 844–862.

Satizábal, P. (2018). The unintended consequences of 'responsible fishing' for small-scale fisheries: Lessons from the Pacific coast of Colombia. *Marine Policy, 89*, 50–57.

Satizábal, P., & Peters, K. (2022). 30% - Re-percenting the ocean. *Presentation at the Royal Geography Society (with IBG) Annual International Conference*, Newcastle, August 30–September 2, 2022.

Steinberg, P. E. (1999). Lines of division, lines of connection: Stewardship in the world ocean. *Geographical Review, 89*(2), 254–264.

Saputra, M. A., & Sammler, K. G. (2024). Volumetric, embodied and geologic geopolitics of the seabed: Offshore tin mining in Indonesia. *Territory, Politics, Governance*, 1–19. https://doi.org/10.1080/21622671.2024.2334821

Smyth, C., & Hanich, Q. (2019). *Large scale marine protected areas: Current status and consideration of socio-economic dimensions.* Retrieved 23 June, 2024, from https://www.pewtrusts.org/-/media/assets/2019/03/mpa-research-agenda-for-public-release-final.pdf

Turner, J., & Peters, K. (2016). *Carceral mobilities. Interrogating movement in incarceration.* Routledge.

Turner, J., & Peters, K. (2017). Rethinking mobility in criminology: Beyond horizontal mobilities of prisoner transportation. *Punishment & Society, 19*(1), 96–114.

Index

K. Peters and J. Turner (eds.), *Ocean Governance (Beyond) Borders*,
https://doi.org/10.1007/978-3-031-71322-4